MET-Vol. 3

INNOVATIONS AND APPLIED RESEARCH IN MECHANICAL ENGINEERING TECHNOLOGY – 2003

presented at

THE 2003 ASME INTERNATIONAL MECHANICAL ENGINEERING CONGRESS
NOVEMBER 15–21, 2003
WASHINGTON, D.C.

sponsored by

THE MECHANICAL ENGINEERING TECHNOLOGY DEPARTMENT HEADS COMMITTEE
OF THE COUNCIL ON EDUCATION, ASME

organized by

SCOTT DANIELSON
ARIZONA STATE UNIVERSITY

THE AMERICAN SOCIETY OF MECHANICAL ENGINEERS
Three Park Avenue / New York, N.Y. 10016

ISBN No. 0-7918-3722-X

FOREWORD

The Mechanical Engineering Technology Departments Heads Committee (METDHC) of the American Society of Mechanical Engineers has a set of three technical sessions within the 2003 ASME Mechanical Engineering Congress (November 15–21, 2003, Washington, D.C.). These sessions help the METDHC meet one of its primary goals, e.g., stimulating and supporting discussions and activities associated with engineering technology education, including student, faculty, and professional matters. Thus in the 2003 Congress, the three sessions scheduled center on applied research projects and laboratory development, innovative teaching techniques and approaches, and administrative issues (promotion and tenure, etc.).

Key individuals who helped the METDHC achieve a successful Congress are Scott Danielson, Program Chair and current chair of the METDHC, Dave Myszka, Past Chair, Theresa Oluwanifise and Marian Heller, ASME Staff, and the large number of individuals fulfilling roles as authors, reviewers, and moderators.

CONTENTS

Proceedings of IMECE'03
2003 ASME International Mechanical Engineering Congress
Washington, D.C., November 15–21, 2003

IMECE2003-42343

A CONTINUOUS IMPROVEMENT / MANAGEMENT-BY-OBJECTIVES APPROACH TO EVALUATING UNIVERSITY FACULTY

Mohammad A. Zahraee, PhD, PE
Purdue University Calumet
Hammond, Indiana
Zahraee@calumet.purdue.edu

ABSTRACT

This paper discusses the design, analysis and implementation of a faculty evaluation system to be used in both departments of Electrical Engineering Technology and Manufacturing Engineering Technologies and Supervision at Purdue University Calumet. The System, based on a faculty member's continuous improvement plan, builds on the Management-by-Objectives approach, which reflects the Human Resource practice of performance plans and evaluations in corporate America. This new system, being outcome based, asks faculty to set goals and objectives with some degree of flexibility and is in line with the accreditation requirement changes of Accreditation Board for Engineering and Technology (ABET).

INTRODUCTION

The new ABET criteria, EC2000 for engineering programs and TC2K for engineering technology programs, is outcome-based. Programs should define their goals and objectives, try to achieve those goals through initiatives, and measure the achievements of their goals through continuous assessment. In addition, having a feedback mechanism should close the loop. This proposed system of faculty evaluation is following the same principal. If programs should have continuous improvement plans, so should the faculty teaching in those programs. This new evaluation system allows faculty to set their own goals and holds them responsible for showing the achievement to those goals. Of course, these goals should be in line with, and in support of, program, departmental, and college goals.

This new evaluation system uses two important documents: the Annual Agreement Regarding Duties and Responsibilities (in which faculty and head agree to allocation of time and resources) and the Annual Evaluation Form (in which both faculty and head evaluate previously and mutually agreed upon goals). The implementation of the new system and faculty acceptance was critical and is also discussed.

HISTORY AND EVOLUTION OF NEW SYSTEM

In the spring of 2000, the head of the Manufacturing Engineering Technologies & Supervision charged a "faculty evaluation committee" to develop a common form for faculty annual review. The committee was charged to develop the form, obtain the buy-in from all other faculty and the department head, and use the process in evaluation of faculty. They were further asked to evaluate all faculty, as an advisory committee to the department head based on the process, and forward their evaluation to the department head. The Committee set a standard for annual reviews, where each faculty had to define a mission statement for himself/herself, set goals in line of the mission statement, and report how his/her activities during the year helped the faculty to achieve those set goals. This evaluation system was a huge jump from the previous model where each faculty used to write an annual report consisting of three areas of Teaching, Research, and Service, and simply report his/her activities during the year in each of these areas.

In the spring of 2002, the Chancellor of Purdue University Calumet asked each department to have a written, documented process for evaluation of all members of the department. At that point, all the faculty of Manufacturing Engineering Technologies and Supervision agreed that the department head should modify the existing process to follow a continuous improvement process. The department agreed that the primary mission should be to design a faculty evaluation process that reflected the value each tenure track or tenured faculty member contributed to the overall success and growth of the department. Other goals included: the development of a process that was perceived as fair and just by faculty; the inclusion of accountability and feedback of the performance for each faculty member; and the development of a standardized evaluation form to rank faculty members.

The initial step of any planned change is to assess the present situation and to identify any deficiencies. First, the existing faculty evaluation system was reviewed. In that system, as explained before, each faculty member was expected to submit an annual activity report outlining his or her goals and professional accomplishments. Although there was no standardized form, the annual report was expected to include: progress toward the accomplishments of professional goals from previous years' annual report, teaching activities and evaluations, scholarship, university and community service, and projected goals for the upcoming year.

The department head then evaluated the annual report. The department head generally used the three areas of teaching, research, and service to establish rankings. The faculty received feedback regarding his or her performance via the next annual contract that reflected the new salary.

This system provided no formal two-way communication between the department head and individual faculty member. It is perceived that the increased communication between the individual faculty member and the department head would increase the ability of the faculty member to understand the goals of the department as well as understand the "value added" of each faculty member in relation to the merit-based pay decision process.

DESIGN AND IMPLEMENTATION OF THE NEW EVALUATION SYSTEM

The efforts at this step were turned on the creation and design of the evaluation instrument, and a process for conducting the faculty appraisals. The evaluation process that has been developed has two major components: the "annual agreement regarding duties and responsibilities" and the "annual faculty evaluation form." (See appendices 1 and 2)

The Annual Agreement is essentially a contract between the faculty member and the department head, which clearly sets forth the criteria that will be used to evaluate the performance of a faculty member and determine merit salary increases. The purpose of this contract is to recognize the diverse interests and abilities of each faculty member as well as to allow the department head to accurately assess whether or not various contracts with the faculty, in aggregate, meet the goals of the department.

The second document focuses on the accountability of each faculty member. In essence, the evaluation form would objectively evaluate the "value added" by each faculty member. This form also provides an exchange of mutual expectations between a faculty member and the department head.

A five level rating scale of excellent, exceeds expectations, meets all expectations, marginal, and unacceptable are to be used to evaluate each of these areas of responsibility. These ratings follow the management performance appraisal of a Fortune 500 company. The proposed instrument also requires a summary rating by both the participating faculty member as well as the department head in assessing the major areas of responsibilities. Similar to a typical performance appraisal in corporate America, a career development section is added to designate if a faculty member does intend to apply for promotion, sabbatical, or grants. The signature section assures that a conference and review has taken place between the faculty member and department head. It should be noted that the signatures are not intended to indicate agreement with the evaluation.

This system was implemented in the spring of 2003 and was used to evaluate twenty two faculty. In general, everybody has been receptive. The effects of this system and how it will help faculty to continuously improve will be discussed at a later paper. The effect of this system on overall continuous improvement of the programs will also be studied at a later time.

CONCLUSIONS AND OBSERVATIONS

The performance plan designed in this paper consists of two major components – The Annual Agreement Regarding Duties and Responsibilities and The Annual Faculty Evaluation. The Annual Agreement is essentially a contract, which clearly sets forth the criteria that will be used to evaluate the performance of a faculty member and determines merit salary increases. The Annual Faculty Evaluation focuses on the accountability of each faculty member in the areas of teaching, discovery, and engagement. This form will be used at the end of the evaluation period and rate the overall performance of the faculty member.

This process was used in the spring of 2003. Although it is a time consuming process, in general, all faculty members seem to be in favor of it. This process follows all other assessment process being done, for quality control, at the program level. The initial results show that a few minor changes need to be made to this document. One change is the period of evaluation. It was quickly observed that with the evaluation being done in March, it is better to have the evaluation period to extend to the end of February instead of the end of the calendar year. By the time the conferences between department head and faculty are done, over 30% of the new evaluation period has expired, not giving enough time to faculty to adjust. The overall effect of this process as well as future changes and observations will be reported in a follow up paper.

REFERENCES

[1] Caruth, Donald L. and Gail D. Handlogten (1997) *Staffing the Contemporary Organizations,* 2nd ed., Westport: Praeger Publishers, pp. 230-231

[2] Schermerhorn, Jr, John R., James G Hunt, and Richard N. Osborn (2000) *Organizational Behavior,* 7th ed., New York: John Wiley and Sons, Inc. p. 400.

[3] Society For Human Resource Management and Personnel Decisions International (2000) *"Performance Management Survey,"*http://www.shrm.org/surveys, Feb. 6, 2001.

[4] Lucas, John, J., Lori S. Feldman, and Philip H. Empey *"A Corporate Human Resource Approach to Evaluating University*

Faculty: Lessons in Development and Communication." Academy of Educational Leadership Journal, Vol. 5, Number 2, 2001.

APPENDIX 1

Department of Manufacturing Engineering Technologies & Supervision

ANNUAL AGREEMENT REGARDING DUTIES AND RESPONSIBILITIES

Name: ____________________________ Rank:____________________________

The following is a statement of expectations based upon value added to the Department regarding your work activities for the upcoming year. The percentage to the right indicates the level of activities agreed upon and the weight that should be given each area in determining merit increases. The percentage of time spent must be at least the minimum for each activity as indicated below. With the Department Head, you will agree to the estimate of time spent for each of the activities listed below. Upon agreement, you and the Department Head will sign this document. A copy of this document will be provided to you. At your annual evaluation, the Department Head will review this document with you and discuss how effectively you have met your mutually agreed upon goals.

I. Teaching Activities: **Weight for Assessment: ____________ (40% minimum)**

You have stated that you would like to contribute to the Department in the area of teaching. Your contribution will be evaluated by the courses taught, course/curriculum development, preparation and/or publication of instructional material, student evaluation, and special activities that have enhanced teaching effectiveness.

II. Publications and Scholarship activities: **Weight for Assessment: ____________ (15% minimum)**

You have stated that you would like to contribute to the Department in the area of scholarship. Your contributions will be evaluated by articles in refereed journals, completed books, articles in non-refereed journals, manuscripts in progress and/or accepted for publication, book reviews, papers presented at technical and professional journals/societies, research grants and awards, and other scholarship that enhance the University's reputation through your scholarship.

III. Service to University Community, and Profession **Weight for Assessment: ____________ (15% minimum)**

You have stated that you would like to contribute to the Department in the area of service to the University, Community, and the Profession. Your contributions will be evaluated by committee service at department, campus, and university levels, student advising, government, industry, or agency service, professional organization service, including dates and offices held, membership in professional societies, leadership role in workshops/conferences, and other activities that enhance the University's reputation through professional service.

The signatures below signify mutually agreed upon activities for the next assessment year.

Faculty Member: **Date:**

Department Head: **Date:**

APPENDIX 2

DEPARTMENT OF MANUFACTURING ENGINEERING TECHNOLOGIES & SUPERVIS ION
Purdue University Calumet

ANNUAL FACULTY EVALUATION FORM
Year: ________

Name of Faculty Member: ________________________________

Academic Rank: ________________________________

Directions:

The intent of this annual faculty evaluation form is to objectively assess the "value added" by each full-time faculty member to the Department.

This form is to be used as follows:

1. The faculty member is to complete appropriate sections and then forward the form to the Department Head by the second Friday of March. The evaluation period will be the "Calendar Year" previous to that March. Faculty members will submit their annual report, students' evaluation of faculty, annual agreement from prior year and other appropriate documents, which will be helpful in consideration of the performance.

2. The Department Head will review the submitted form and supporting documents. Additionally, the Department Head shall complete appropriate sections, and thereafter shall schedule a conference with the faculty member, at which time the completed form will be discussed. This conference should take place before mid-April.

3. At the conclusion of this conference, The faculty member will be asked to sign the form. The faculty member's signature signifies that he/she has discussed their performance with the Department Head and has also seen the Department Head's comments but not necessarily that he/she agrees with all of them. The Annual Agreement regarding Duties and Responsibilities will also be discussed and signed for the next assessment year. The faculty member may respond to the department head's evaluation in writing no later than 2 weeks from the date of his/her evaluation conference.

4. One copy of the evaluation form will be given to the faculty member and one form will be placed in the faculty member's personnel file.

Disclaimer: This document and process is used only for performance evaluation and career development. This evaluation instrument is not intended to be used as part of the tenure process.

Teaching Ability and Effectiveness
Potential subjects for assessment: Number and levels of Courses taught, student evaluation, contributions in Course/curriculum development, preparation and/or publication of instructional materials, and special activities that have enhanced teaching effectiveness. More activities could be initiated and included by faculty member.

Faculty Member's Assessment/Comments

Overall Rating For Teaching

_________Excellent
_________Exceeds Expectations
_________Meets all Expectations
_________Marginal
_________Unacceptable

Department Head's Assessment/Comments

Overall Rating For Teaching

_________Excellent
_________Exceeds Expectations
_________Meets all Expectations
_________Marginal
_________Unacceptable

Publication & Scholarship Activities

Potential subjects for assessment: articles in refereed journals, completed books, articles in non-refereed journals, manuscripts in progress and/or accepted for publication, book reviews, papers presented at technical or professional meetings, editorship of professional journals, and scholarly grants and awards. More activities could be initiated and included by faculty member.

Faculty Member's Assessment/Comments

Overall Rating For Scholarship

_________Excellent
_________Exceeds Expectations
_________Meets all Expectations
_________Marginal
_________Unacceptable

Department Head's Assessment/Comments

Overall Rating For Scholarship

_________Excellent
_________Exceeds Expectations
_________Meets all Expectations
_________Marginal
_________Unacceptable

Service to the University, Community, and the Profession

Potential subjects for assessment: Committee service at campus, department, and university levels, student advising, government, industry, or agency service, professional organization service, including dates and offices held, membership in professional societies, leadership role in workshops/conferences, and other activities that enhance the university's reputation through service.
More activities could be initiated and included by faculty member.

Faculty Member's Assessment/Comments

Overall Rating For Service

_________Excellent
_________Exceeds Expectations
_________Meets all Expectations
_________Marginal
_________Unacceptable

Department Head's Assessment/Comments

Overall Rating For Service

_________Excellent
_________Exceeds Expectations
_________Meets all Expectations
_________Marginal
_________Unacceptable

Performance Evaluation for Release Time

Please check the appropriate categories listed below:

_______Advising Release
_______Administrative Release
_______Scholarly Release
_______Other

Overall Rating by Faculty

_________Excellent
_________Exceeds Expectations
_________Meets all Expectations
_________Marginal
_________Unacceptable

Performance Evaluation for Release Time

Please check the appropriate categories listed below:

_______Advising Release
_______Administrative Release
_______Scholarly Release
_______Other

Overall Rating by Dept. Head

_________Excellent
_________Exceeds Expectations
_________Meets all Expectations
_________Marginal
_________Unacceptable

Career Development

Please check the appropriate Categories Listed below:

_________Intends to apply for promotion during academic year
_________Intends to apply for sabbatical leave
_________Intends to apply for PRF International Travel Grant

The signatures below indicate that a conference and review of performance has taken place

Faculty Member	Department Head
Printed Name:	***Printed Name:***
Signed Name:	**Signed Name:**
Date:	**Date:**

Proceedings of IMECE'03
2003 ASME International Mechanical Engineering Congress
Washington, D.C., November 15–21, 2003

IMECE2003-42360

INNOVATIVE APPLIED RESEARCH PROJECTS USING INDUSTRY COLLABORATION

Saeed Foroudastan, PhD/Middle Tennessee State University, Engineering Technology Department

Linda Hardymon/Middle Tennessee State University, Center for Energy Efficiency

KEYWORDS

Applied, collaboration, laboratory, industry, partnerships

ABSTRACT

Partnerships between the Middle Tennessee State University Engineering Technology Department and the local industrial community are well adapted to research and development projects for the students. Resulting interactions between engineering technology programs and industry are advanced in many ways, including long-term partnerships, informal contacts between faculty members and industrial personnel, consulting, and collaboration on training opportunities, discussions, seminars, and teaching programs. Foremost, however, are the many ways students benefit from the related assignments. Through applied research projects, students focus on innovative project developments that provide practical solutions to complex problems. They learn to initiate, design, and implement new initiatives within the university and industrial partnerships and to profit from the opportunities to explore new technologies and practice skills that meet real world challenges.

MTSU modified its introductory engineering course to incorporate not only the fundamentals required to the support basic engineering course learning experience, but to include a choice of hands-on design projects. At present, a solar powered vehicle and a moon buggy have been designed, constructed, and entered into national races to test the decisions and capabilities of the projects as a result of this innovative laboratory approach. The laboratory atmosphere centered on these applied research projects features the opportunity to work at the university and/or take advantage of the industrial partner's equipment and expertise.

Partnerships with industry are essential in providing access to the latest equipment and technology. Applied research projects are important for students to gain a much better "sense" of engineering and to progress to higher levels of project interaction that offer design and design problem issues, use of knowledge, physical application, and comprehension of engineering principles.

INTRODUCTION

At Middle Tennessee State University (MTSU), the Engineering Technology and Industrial Studies Department (ETIS) strives to provide the best educational opportunities available for students majoring in engineering and engineering technology fields [1]. Some courses employ a combination of instruction and application using creative options within the program that encourage, challenge, and stimulate interest. An introductory engineering course that encourages freshman to be involved with the topics being introduced kindles interest by incorporating a design project in the course. Using a combination of general educational basics in the applied sciences plus outcome-based applications, the students develop class projects that merge the technologies being introduced and real-life use of the information. For the concept to be successful, the use of state of the art equipment and materials are often needed. For that reason, the projects selected take advantage of partnerships in place with numerous local industries to pool resources for completion, gaining the students the advantage of a relationship in the industrial sector. The sharing of resources has also had a positive effect on upper division courses and capstone projects.

The ventures encourage interest in the academic program, aid in retention [2], and prepare students for engineering activities in the workplace. Project integration in the courses provides students with an opportunity to see the relationship between their lecture and class work activity and real life ever-changing technologies. The addition of projects also serves to demonstrate the connections among engineering and technological applications, teamwork and communication abilities, and analytical skills. It is important that students be encouraged to apply that knowledge. This increased contact between a student and real world opportunities as provided by the use of equipment and advice from industry partners is important.

With troublesome budget issues on the rise, however, funding constraints limit the presence or development of technology laboratories on campus with equipment sufficient to support certain projects, and, thus, the prospects for applied or useful exploration. The availability of specialized equipment required for some projects is a limiting factor. That is why partnerships with industry are an essential avenue in providing access to the latest equipment and technologies. Applied research and development projects are important for students to gain a much better "sense" of engineering and to advance to higher levels of project interaction with design and design problem issues, use of knowledge, physical application, or comprehension of engineering principles. Collaborative efforts between MTSU and its partners in industry [3] provide access to the latest technologies in the workplace and hands-on work opportunities. Justification for the expense of supplying a laboratory with modern equipment and the training of instructors is difficult, however, joint use of an industry partner's equipment together with equipment currently available on campus permits an applied research laboratory to exist. Instead of building and supplying a laboratory space, a laboratory atmosphere is built around the projects using industry equipment as needed. Innovative ideas such as this that can aid a university in providing a leading-edge education must be explored. A laboratory effort that joins industry and academia in providing access to the latest technology available prepares graduates for jobs and meets that need.

PROJECTS

MTSU first modified its introductory engineering course to include not only the fundamentals required to support basic engineering course learning experiences, but to introduce a choice of design projects. The hands-on design idea made the course more interesting and more challenging, and gave the students a taste of real-world applications as motivation to continue with engineering or engineering technology as their major. Through the first design project, which initiated the concept of project involvement by way of simple drawings and models, the students gained a better understanding of their major as a career choice because of the exposure to the various engineering disciplines the project required. Since that time, a solar/battery-powered vehicle "Fig.1" has been designed and constructed with support from the university and industry. A bonus to the project was the opportunity to enter in an organized competition at a national race to prove the capabilities of all decisions made regarding the students' and team's design and construction choices.

Fig.1: 2002 Solar vehicle at the race

Overall, the project that began as a simple design problem and developed into an entry into a national race has been a huge success. It has offered problem-solving, communication, project management [4], and ethics as educational advancements, and provided the students with a better 'sense' of engineering as a career. The vehicle is currently undergoing modifications as a class project "Fig.2" for entry in a national race this year. Design and casting of a polyurethane body for the vehicle using advice and expertise from industry is the latest collaborative laboratory effort in that project.

Fig. 2: 2003 Solar vehicle awaiting body and solar panels

As a result of the participatory relationship with industry from the solar project, another was undertaken. A NASA-sponsored competition involving design and construction of a vehicle for an outer space application, a moon buggy, was entered. Following specific design and construction regulations for entry in the national competition, the moon buggy "Fig. 3-6" was designed by the students and fabricated with assistance from local industry. Using joint access to equipment, the shared research laboratory structure that provided the tools and assistance necessary to take the step from concept to construction, the students were able to compete with top engineering and engineering technology schools at the NASA event and achieve a high degree of personal accomplishment by finishing third overall. This success was a direct result of the collaboration with industry and the joint laboratory effort.

Fig. 3: Overseeing moon buggy construction

Fig. 4: 2003 moon buggy at the NASA competition

Factors for project selection for the courses are simple. Each project must complement the fundamental engineering education for the course involved, agree with the learning objectives of the class, and provide an understanding of engineering disciplines. Projects must be do-able and short-term with plans for the project to be accomplished, whether the plan is to design only, design and construct, construct from a previous plan, or improve on an existing design, and be finalized in one semester. To add excitement and stimulus, since there are many opportunities available, the MTSU project selection is based on an expectation to participate in a national competition in order to validate the project, to compete against other universities (sharing and learning from their experiences), to make the investment of time and effort more valuable, and to increase the stakes in a job well done.

Fig. 5: Moon buggy flying over the course

Project selection must be results oriented, offer positive learning experiences to engage the students in the subject matter, and encourage the students to take additional engineering or engineering technology classes. For the two projects mentioned, many students have remained connected to the projects after the semester has ended through student organizations. Of the students involved, about ninety-percent are or have become members of the student chapter of the American Society of Mechanical Engineers (ASME), which has also helped sponsor the entry into the national competitions. To achieve the goal of student interest and involvement, the collaborative laboratory concept must be built around such innovative projects.

Fig. 6: Proud moon buggy team 2003

INDIVIDUAL ACCOMPLISHMENTS

Student teamwork and hands-on activity resulting from the projects and the affiliations add value to all involved. For each project, a hierarchy is established under the administrating professor with a team leader and each participant having specific responsibilities to keep the project organized and on target.

The solar bike project included several sub-teams. For example, the electronic and battery group studied horsepower and gear ratios, calculated amperage usage strategies in order to complete the thirty-four-lap race within the three-hour time limit, and researched the best equipment available before purchasing to stay within their budget. Additionally, devices were mounted in the vehicle to monitor speed, amps spent per hour, and battery usage. Gear ratios were varied to make the most of solar and battery power.

Others sub-teams included motor and controller, steering system and brakes, chassis construction, and solar technology. For the chassis and bodywork, result evaluations were based on aerodynamics, coefficients of drag, and technology factors important to the construction. Machining parts, welding, metal forming, and molding were included. An oven was built to fabricate the vehicle's body shell and canopy. Consultations with industry on some aspects of the project were essential for a successful project and rewarding experience.

The moon buggy project team had a similar arrangement with groups responsible for design, metal forming and machining, gearing and chain mechanisms, brake system, assembly, and others, all adhering to competition regulations. Construction issues were major concerns. In particular, the students realized converting a classroom CAD project to an actual construction project was a challenge requiring several refinements and revisions in order to fabricate the parts in the machine shop.

The national competitions with other universities gave the students self-confidence as a team and as individuals. Now, when seeking a future in an industrial, manufacturing, or engineering field, the students have a greater level of self-assurance and accomplishment because of the projects, the teamwork, and the collaboration with others.

THE CONCEPT

In some cases, laboratories are created with an eye toward helping business and industry investigate and resolve problems of a technical nature. The creation of an applied research laboratory at MTSU to offer assistance with the types of projects undertaken is not easy, so the alternative is to partner with local industries, using their state of the art facilities, equipment, and technical expertise to provide the students with up to date – 21st century – learning opportunities. This innovative approach is not limited to one piece of equipment at one facility, but allows access to different equipment at different facilities that meet the needs of the individual students and their particular projects. The more a student works with industry, the more they learn about the demands of real world applications. Faculty members are encouraged to take active roles with the students in the use of equipment and development of the lab opportunities. Cooperative study opportunities have resulted from the relationships. Taking advantage of local industry interests in developing relationships with a university focus and presence is a valuable experience for all involved.

The applied laboratory concept at MTSU is centered on, not building a laboratory and completing a project, but developing access to a laboratory environment for the applied research projects through a joint effort with partners within industry. It features the opportunity to work and study at the university as well as take advantage of the industrial partner's equipment and expertise. The collaborative effort motivates students, improves their performance, and allows them the use of technologies otherwise unavailable to them.

PARTNERSHIPS

Partnerships between the MTSU Engineering Technology Department and the local industrial community already exist and are well adapted to research and development projects for the students. Resulting interactions between engineering technology programs and industry are advanced in many ways, including long-term partnerships, informal contacts between faculty members and industrial personnel, consulting, and collaboration on training opportunities, discussions, seminars, and teaching programs.

Foremost, however, are the many ways students are encouraged to develop projects, write reports, and participate in related research assignments. Through applied research projects, students focus on innovative project developments that provide practical solutions to complex problems. They learn to originate, design, and implement new initiatives within the structure of the university and industrial partnerships. They profit from the many opportunities to explore new technologies and to practice skills that meet real world challenges.

The plan to develop a more intense exchange of information and equipment use that benefit students has barely begun. However, with cooperation and collaboration, not only the students, the faculty, and the University benefit, but involved local industry partners benefit as well. As an industrial or manufacturing community of skilled employees with a willingness to share their experiences and expertise with the students is exposed to a university focus, many of those employees are encouraged to continue with their educations.

Aspirations are for the partnerships to supply applied laboratory experience for the students and include expanding partnerships between MTSU and the industries in the area, pursuing efforts to continue more collaborations with other industries, providing an atmosphere allowing for achievement and personal growth for students and faculty, and developing outstanding students. The partnerships have also brought some degree of financial assistance to the University.

CONCLUSION

After a preliminary survey was conducted following completion of an earlier project, responses point out the level of excitement among the students. Those who had participated in a design project had a higher level of enthusiasm for their major. Approximately eighty-five-percent indicated they were encouraged to continue in engineering and engineering technology courses as a result of the hands-on experience. Others indicated they were made more aware of the multiple career directions for engineering and engineering technology majors due to the participation with actual projects and access to industry. Overall, the hands-on project experience gave the students a better understanding of what the major offered and felt encouraged to pursue it as a career. From the application of the project to the use of knowledge, from the activities involved with the project to completion of the project, students felt they faced an attractive career choice.

Since typical project assignments are introduced at later class levels, the use of this hands-on opportunity encourages interest early in the program and provides exploration of the career potential for degrees earned. Without the collaborative laboratory effort, students are not as engaged in thinking and problem solving applicable to real-life situations, and their interest and enthusiasm are more difficult to maintain.

Successful completion of the solar vehicle and the moon buggy projects, along with success in the national competitions, and student enthusiasm are proof that this is a winning situation for MTSU. With shared equipment opportunities, use of equipment at the University and at the industry site as required, and benefit of the advice and expertise of those working in industry, an applied research and development laboratory centered on applied research projects for the students provides winning outcomes. It is the most cost effective means for MTSU to gain access to an applied laboratory atmosphere.

REFERENCES

[1] Akins, L., "Partners in Recruitment and Retention", 2001 Proceeding of the ASEE Annual Conference and Exposition, Albuquerque, NM, June 24-27.

[2] Hirsch, P., "Enriching Freshman Design Through Collaboration with Professional Designers", 2002 Proceeding of the ASEE Annual Conference and Exposition, Montreal, Quebec, Canada, June 16-19.

[3] Foroudastan, S., et al, " Partnering with Industry – A Winning Collaboration", 2003 CIEC Annual Conference, Tucson, Arizona, January 28-31.

[4] Vavreck, A. N., "Project Management Applied to Student Design Projects", 2002 ASEE Annual Conference and Exposition, Montreal, Quebec, Canada, June 16-19.

Proceedings of IMECE'03
2003 ASME International Mechanical Engineering Congress
Washington, D.C., November 15–21, 2003

IMECE2003-42364

Using Videotape to Add an Asynchronous Delivery Option to Regular Classroom Instruction

John W. Blake, Ph.D., P.E.
Associate Professor and Chair
Department of Engineering Technology
Austin Peay State University
P.O. Box 4536
Clarksville, TN 37044
USA
(931) 221-1476
FAX: (931) 221-1493
blakej@apsu.edu

ABSTRACT
Student schedules are often at odds, forcing them to miss needed classes due to time conflicts. Departments have limited faculty resources, and cannot offer courses every term at times to meet every need. This paper describes the use of instruction via videotape to offer students an option for taking courses when they cannot attend at the scheduled time. Delivery of instruction by videotape is not new; this paper describes efforts to adapt this method to specific circumstances and student needs. This has resulted in higher course enrollments when videotape and traditional sections are combined for teaching load purposes, and has permitted students to complete graduation requirements when, without this option, they would have been delayed due to scheduling conflicts.

INTRODUCTION
Courses by videotape and other distance learning technologies serve people who cannot attend traditional courses (1,2). Videotaped courses have been available for many years (3) at both the undergraduate and graduate level from many universities (4, 5). Schools with major distance learning programs designed to attract a national audience dedicate significant resources, such as the assistance of instructional designers and experts in distance learning technology, to develop and support their programs (1).

This paper will address a different use of videotape instruction. Our department is not attempting to offer a degree through distance learning. Rather, we have augmented our campus-based program by offering video sections in parallel with traditional classroom courses. These meet the needs of students who cannot attend a course at the scheduled time. This is done with very limited resources.

Our Engineering Technology program operates at a satellite campus located on Fort Campbell, KY. We are part of Tennessee's designated comprehensive liberal arts university, and exist to meet regional needs for a four-year degree program in technology. Many of our students are nontraditional, including active duty military, retirees, and civilians from the local community. Our area has a number of manufacturing plants, and many of our students work for those industries. Most courses are offered between 4:45 and 10:00 PM on an accelerated eight-week schedule. Three semester hour courses meet twice a week for two and a half hours per class. Missing one class is the equivalent of missing an entire week on a standard sixteen-week semester. Daytime offerings are limited, which limits options for students who work second shift.

Each term, we have students who face time conflicts. Some students need two courses that are scheduled for the same time. With the long class meetings, this problem is worse here than with traditional sixteen-week semesters. Some students need classes that conflict with their work schedule. Some students must miss some classes due to training or travel required by their workplace. Despite the conflicts, our mission is to serve these students.

For some courses, we offer an asynchronous videotape option to serve these students. The regular class is taped, and students who cannot attend the regular sessions take the course by video. This has made it possible for students to get required courses despite scheduling conflicts and to graduate on time. As the video section allows more students to take a course, this also boosts the department's student credit hour production.

DEVELOPMENT OF THE VIDEOTAPE OPTION

The department's practice of offering courses by videotape began in Fall 2000, with a student who needed one last course for graduation. The course was offered once a year, and the scheduled time conflicted with the student's work schedule. He could not graduate until he completed this course. Waiting for next year was not an acceptable option. The solution was to videotape each lecture for the student. I had done this before when I knew a student was to be absent, but had not attempted to videotape an entire course.

In most terms since that time, I have had at least a few students taking a course by videotape. I have offered our introduction to engineering technology, problem solving, materials, thermodynamics, and machine design courses by video. All are currently taught as traditional lecture and problem-solving courses. At times, the video program has grown to where enrollment in video sections has rivaled enrollment in the regular section. With some marginal courses, the video section provided badly needed additional enrollment.

PRODUCING PARALLEL VIDEOTAPE COURSE SECTIONS

Each class is taped using a regular 8mm video camera fixed on a tripod. The result of this low budget production is a tape that gives students the essential elements from the class. The department does not have a studio classroom; the equipment is portable and is taken to the assigned room for each meeting. The instructor operates the camera, and must be careful to keep within the camera's field of view. The lecture is interrupted when a new tape is needed. Usually, the class is given their break (one break during a 150 minute class period) at one of the tape changes.

The course is taped each time a regular section is offered. While I could use tapes from past terms for video students, I do not consider any of my courses to be static and want the video students to get the latest version of the course. I can and do use old tapes if a student must take a last course for graduation when I am not offering a regular section. This is an improvement over our department's past practice of offering students an independent study option in this special case.

At first, I set the camera to cover a large section of the marker board and the lectern. Students reported that they could not read the board at this scale, and requested that I zoom in on a smaller section of the marker board. With this arrangement, students get a close view of a limited section of the board. This has two drawbacks: the instructor is restricted to a small working area and the students can only hear the disembodied voice of the professor. Despite the drawbacks, this has worked relatively well.

After each class, the 8 mm tape must be copied to VHS format, and then copies made for each student. Again, this is done by the instructor. This process takes time, and causes the video section to lag behind the regular class. I allow for a time lag of one week in the schedule. In the past, I have put one week's worth of classes onto each VHS tape. This required fewer tapes, but it took more time to get tapes to students. I was able to produce tapes faster when I issued a separate tape for each lecture. This did require more tapes. While students are expected to supply tapes, the department has also had to supply tapes on occasion, which can become costly. Students are encouraged to return tapes after the term ends so that they can be recycled.

We recently replaced the older video camera. The new camera has a side view panel that can be placed so that the instructor can see what is being taped. This has been a great improvement and was worth the extra cost. If I had a VCR in the classroom, I could make the VHS tapes directly during the class. As this would require me to bring more equipment, I have not done so.

STUDENT ISSUES IN PARALLEL VIDEOTAPE COURSE SECTIONS

Students taking the video course register for a separate section and receive a different syllabus. They are given special instructions on contact with the instructor and different test dates. The list of special instructions has grown with experience. Due to the time lag, they are told to expect an incomplete at the end of the term.

Students must have e-mail and access to a VCR. I make and use an e-mail distribution list for class communication. General announcements are sent to the group. I use individual addresses to send exam scores and other personal communications. Students are instructed to check their e-mail at least twice a week. I expect to have contact – by e-mail, telephone, or in person - with each student at least once a week.

Some students approach a video course with trepidation. These students usually do very well. They come regularly to pick up their tapes, make time to watch the tapes each week, and spend time with me to discuss questions. Sadly, others do not approach the course with the proper attitude. These students come in sporadically to pick up tapes, and take very little advantage of opportunities to see the instructor for help. Often, these students do very poorly. Students often take a video section along with regular courses, and the regular course demands are given priority. This can lead to academic disaster in the video course; with an eight-week term, one has very little time to catch up.

Good student practices in the video section are similar to good student practices in a regular class. Likewise, there are similarities in poor student practices. The effects of bad practices appear to be magnified in video sections. Course policies and requirements are designed to help students avoid bad practices and to be successful.

Students are required to stay in contact and to sign for their tapes. This takes the place of recording attendance in the regular class as required for some federal aid programs. Students are required to make initial contact by the end of the first week. Early in the term, we are required to report any student in regular classes who registered but never attended. A student who fails to make contact in the first week is reported similarly. Before setting this policy, I had one extreme case where a student did not make contact until the third week of the eight-week term, and only did so after being tracked down in other classes. The student had a heavy load, and the video course was never given priority. The course policy also states that a student who fails to make contact for more than two consecutive weeks may fail the course due to attendance.

Students sign for and pick up tapes near the department office. Any class handouts are put out along with the tapes. A tape is made for each student. While I encourage students to watch the videos in a group, most must watch individually. A drawback of video instruction is that students do not have the same opportunity for group interaction. Viewing in a group gives them some opportunity to discuss with other students.

Another drawback of video instruction is that students are not able to interact with the instructor in real time. Students are told to make written notes of questions and, to help give me the context, the point in the lecture when they wanted to ask the question. They then e-mail, call, or see me about the question. I find that e-mail works best for simple questions, and that face-to-face meetings are necessary for more difficult issues. Questions answered by e-mail are often similar to those that can be answered directly in class. Often, I send answers to the entire group. Questions requiring office meetings would often require a meeting for students in a regular classroom section. When possible, I encourage students to come to my office in a group so that each can benefit from the questions of others. This also makes better use of my time.

I require the same homework of both regular and video sections. After the homework is due, I give out solutions as feedback. With both groups, this has the risk of students copying from handouts in circulation instead of learning by doing the homework. A student who copies from handouts will not do well on the exams.

With the one-week delay required for copying tapes, examinations for the video group are scheduled one week after those for the regular section. Students come in to the office at their convenience and take the exam. If the student must come in after hours, I either stay late or arrange for them to sit in with another class. The time lag raises some concerns for test security. I usually give similar but different exams to the two groups. Students in the regular class are reminded of the video section and instructed not to discuss the exams outside of class. I use the same scale for scores from both sections. As I often give exams where I need to adjust the scale, students in the first group could hurt themselves by giving out information to the second group that gave them higher scores. This discourages the first group from giving information to the second.

After exams are graded, scores are sent individually by e-mail. Students must come to see me (or, if I am out, the department secretary) to pick up their exams. I expect each student to see me to go over the exam.

With the one-week time lag, grades are due before the video students can view the last week of classes and take their final exam. Grades of incomplete are entered to meet the deadline, and are changed once final grades are available.

CHALLENGES OF VIDEOTAPE COURSE SECTIONS

Video sections present additional challenges for both students and instructor. The student does not have the structure of regular class meetings. With our operating schedule, they must have the self-discipline to make themselves watch five hours of class on videotape each week. This is even more difficult for students with families who are attempting to watch the lectures at home. Students do not have the usual opportunities for interaction with other students or with the instructor. Extra effort is required for both instructor and students to have the necessary interaction. The instructor must be prepared to spend a significant amount of office time helping students. Both students and instructor must find times to meet that are mutually acceptable.

There are drawbacks to taking the class by video. The marker board is harder to read on the video, and the student is not in the room to ask for clarification. I take care to say what I am writing; students report that this helps. Students report that some marker colors that are acceptable in the room are not easy to read on video; black is the only color that seems to be acceptable. Markers must be fresh. A marker that may seem to the instructor to be useable may not produce legible writing on the videotape. I have resorted to putting dates on markers and replacing them regularly.

There are also advantages to classes by videotape. I joke that students can experience my lectures with fast forward and mute options. I have often had students respond by pointing out that they also have rewind and pause options; apparently, they find these to be very useful. If they have trouble understanding part of the material, they can go through that section again. This would be more useful than just an audiotape of the lecture.

COMPARISON OF OUTCOMES BETWEEN STUDENTS TAKING THE COURSE BY VIDEO AND IN THE REGULAR CLASSROOM ENVIRONMENT

As with any alternative to regular classroom instruction, it is of interest to compare student outcomes. As similar exams and grading structures were used in both sections, one can compare outcomes by comparing final grades.

Final grades are compared here for two courses: materials science (Fig. 1) and thermodynamics (Fig. 2). Both are junior level courses in our program. No clear bias towards either classroom or video sections appears in either comparison.

Figure 1: Final Grade Comparison for ENGT 3000 Materials Science

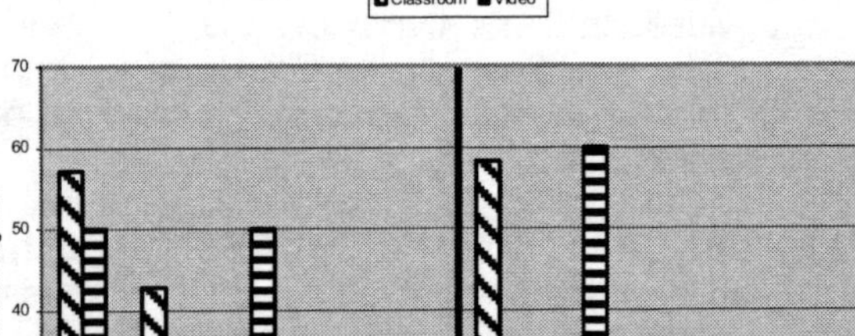

The materials course is a lecture class. For the terms shown, the exams were take-home tests requiring short written answers. These were long and tedious, but the take-home format gave students ample time to find answers. This resulted in a distribution skewed towards grades of "A" and "B" in both classroom and video sections.

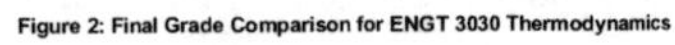

Figure 2: Final Grade Comparison for ENGT 3030 Thermodynamics

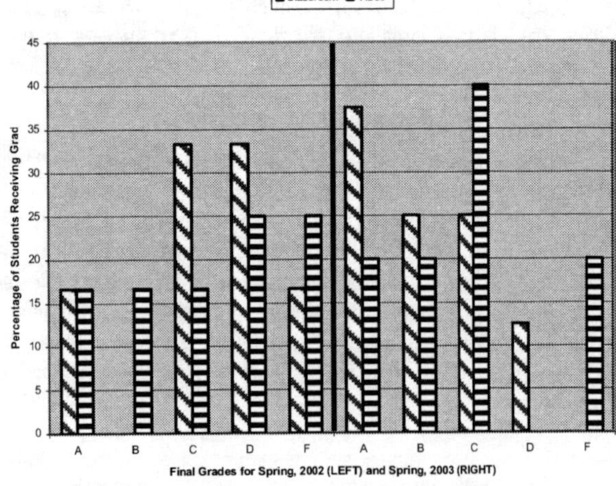

The thermodynamics class is a problem-solving course, and is by far the harder of the two courses as taught. While no bias appears, it is noteworthy that a significant number of video students performed better than students in the regular class.

In my observation of students attempting to take courses by video, I have observed two groups. One group is wary of video courses, and takes the course very seriously. They are prompt in getting and watching videos and in seeing me when they have questions. In general, this group does well. The other group does not give the video course a high priority. They fall behind on the videos and do not come in with questions. These students do poorly in the course. Those in the second group often are attempting to do too much; their work demands, regular classroom courses, and family commitments do not leave them with enough time for the video class.

With rare exceptions, students who have taken a course by video tell me that they would prefer traditional classroom instruction. As the instructor, I prefer the traditional classroom as well, and would only recommend the video option if the student could not attend the class as scheduled. It is noteworthy that, after taking one course by video, many students have taken other courses by video as well.

FUTURE DEVELOPMENTS

There is room for improvement in our video program. The videotape option program described here was developed out of necessity. I did not have assistance from instructional design experts, and other duties have limited my ability to study the best practices of others to apply in this program. Direct feedback from students has been used to improve our program; other avenues for improvement could be found in the experiences of others. This has only been used for lecture courses as designed for classroom instruction. It could be used for a course with lecture and laboratory components for the lecture material; students and instructors would have to find mutually agreeable times to meet for the laboratory section.

To date, I have used videotape to disseminate the regular lectures. For problem-solving classes such as thermodynamics, I find my time is limited for solving examples and homework problems, and I give out printed solutions for students to study. Some students report that it would be more helpful to watch me solve the problem. If time permits, it would be useful for me to tape problem-solving sessions in addition to the regular lectures. These tapes could be borrowed and viewed by students who need additional help with the material.

DVD technology should be explored as an alternative to videotape. Some students have DVD players but must purchase a VCR for the course. While the current price of VCRs is often less than the textbook, it would be desirable to move to the newer format.

One weakness of the video courses described here is that they follow the traditional lecture format. They do not include group discussion, team projects, or student presentations. Technology developed for on-line courses could be used to great advantage with video instruction. A course web site will be added for future video courses. Discussion boards or chat rooms should be added to allow interaction between students. With our current practice, this interaction is sorely lacking. Instead of requiring students to come to campus for exams, they could take their exams over the web.

While it is technically possible to transfer video over the web, current limits on speed and bandwidth make this impractical for sending full class sessions. In the near future, this is likely to become feasible, and the video course can become a web course. In addition to video-lectures from the instructor, student presentations could be videotaped and distributed to the class as well.

CONCLUSIONS

Necessity drove the development of a video course option at Austin Peay. This has evolved as a low budget supplement to regular classroom instruction. The video option has allowed students to complete required courses despite time conflicts, has made courses available to more students, and has added to department student credit hour production. To date, we have not developed a standard survey instrument specifically for video sections. Discussion with students has indicated that this is more challenging than a regular classroom course. A few will not attempt a second video course. Usually, this is due to either a need for the regular class structure or difficulties in taking a course at home. Many students have returned to take other video courses. While it should not be the first choice over classroom instruction, it is a viable option that benefits both student and department.

ACKNOWLEDGMENTS

I want to thank the first video student – Steve Smith – and all who followed him for their help in developing the program described here.

REFERENCES

1. George, Tracy, 2001, A Wired Education, IIE Solutions, **33**, pg 24.
2. Ryan, Margaret, 1997, Education Casts a Wide Net, Electronic Engineering Times, **978**, p 335.
3. Anon, 1980, Videotape Courses Available to Engineers, Design News, **36**, pg 116.
4. Learnon.org Continuing Engineering Education Web Page, American Society for Engineering Education.
5. Anon, 2001, Engineering, U.S. News and World Report, Oct 15, 2001, pg 64.

Proceedings of IMECE'03
2003 ASME International Mechanical Engineering Congress
Washington, D.C., November 15–21, 2003

IMECE2003-41490

HEAT TRANSFER BETWEEN PARTICULATES AND SURFACE IN A VIBROFLUIDIZED LAYER

Vladimir Sheyman, PhD
Division of Engineering Technology
College of Engineering
Wayne state University
Detroit, MI. 48202

Mulchand S. Rathod, PhD, PE
Division of Engineering Technology
College of Engineering
Wayne state University
Detroit, MI. 48202

ABSTRACT

The apparatus, in which fluidization of the particulates is obtained by filtration of the gas in the upward direction through the layer of the particulates, is widely used in the industry for various types of processes. This gas serves as source of fluidized agent and the heat transfer carrier. In particular, the fluidized bed creates a favorable condition to carry out numerous heterogeneous processes. In spite of a significant advantage of this hydrodynamics regime, there are some drawbacks. The fluidization occurs only when sufficient filtration velocity of gases is achieved. In some cases when there is a significant difference in particulate sizes, this condition is not acceptable, and for a layer of fine particulates, such type of fluidization cannot be obtained at all. In the latter case, a combination of vibration and filtration of very small amount of gas through the layer is used, and the heat transfer to the solids occurs by conduction from the heating surface.

In this paper, a set of differential equations for the heat and mass transfer are formulated and solved for variable and constant temperature of the heating surface. As a result, a relationship of the solid temperatures as a function of time is established. An experimental research also was performed. A comparison of the experimental research data and theoretical analysis shows that the deviation does not exceed 10%. The experimental research also shows that the main influence on the intensity of the mass transfer process has the amplitude of vibration and to the lesser extent the effects of frequency of vibration.

NOMENCLATURE

A*	Amplitude of Vibration, mm
C_p	Specific Heat, kJ/kg ^{0}C
f	Frequency of Vibration, Hz
F	Heating Surface Area, m^2
G	Mass of Solids in the Container, kg
h	Height of Dense Layer of Solids in the Container, mm, or Convection Heat Transfer Coefficient, W/m^2 ^{0}C
r	Latent Heat of Evaporation, kJ/kg
t	Temperature of the Heating Surface, ^{0}C
X	Moisture Content of Solids, kg of moist/kg of dry material
?	Mean Temperature of Solids, ^{0}C
t	Time, min
?	Relative Coefficient of Drying

Subscripts

cr	Critical Value
ef	Effective
eq	Equilibrium
f	Final
l	Liquid
o	Initial Value
s	Solids

INTRODUCTION

The apparatus, in which fluidization of the solid partic ulates is obtained by filtration of the gas in upward direction through a layer of the particulates, is widely used in the industry for a number and types of processes. In particular, the fluidized bed creates a favorable condition to carry out various heterogeneous processes. In spite of significant advantages of this hydrodynamics regime, there are some drawbacks. Normal fluidization occurs only when sufficiently and uniformly distributed filtration velocity of gases is achieved. In some cases, when there is a significant difference in particulate sizes, these conditions are difficult to achieve, and for the layer of fine particulates such type of fluidization cannot be obtained at all.

The fluidization of a fine particulate layer is easy to obtain by applying a combination of gas filtration through the layers and vibration. The vibrofluidization is also useful when there is a wide range of particulate sizes. In this case, the fluidization is obtained under relatively small filtration velocity of gases to prevent carrying away of the smallest fractions of particulates from the apparatus. This method is especially useful for thermal treatment of very fine adhering solid materials that cannot be easily discharged from the apparatus because of their adhesiveness. In a number of cases, it is useful to supply heat to the vibrofluidized bed by conduction from the heated surface, especially when the direct contact of the material with the gaseous heat carrier is not acceptable. This method of heat supply leads to a significant reduction of the amount of the filtrating gaseous heat carriers, which is important for the thermal treatment of the very fine particulates.

The heat transfer in case under consideration, when temperature of the surface is constant, is well known in literature [1]. However, the thermal conduction heat transfer from the constant temperature surface to the layer of particulates has limited applications. In practical applications, heat is transferred to the surface by superheated vapor, electric heater, induction heating, hot gases, etc. In these cases, the temperature of heating surface may be constant or variable. To maintain the constant temperature of the heating surface, the amount of heat supply should very in time, which is practically not easy to achieve. Thus, it is important to consider the heat transfer process for a generalized case when the temperature of heating surface is variable. Then, a particular case for constant surface temperature may be derived from the generalized treatment. In this paper, the theoretical and experimental methodologies are applied to address this situation.

THEORETICAL ANALYSIS

As an example of the thermal treatment of powdery materials, an application to heat and mass transfer processes and in particular the drying process, is addressed. The kinetics of heating the moist materials is carried out for a vibrofluidized layer of particulates when heat is supplied to the layer from the heated surface.

The following assumptions are made [1,2]: the temperature of heated material is the same over the height of vibrofluidized layer at any particular time; there exists no gradients in heating of the particulates; there is no variability of the effective heat transfer coefficient from the heating surface to the layer, and this coefficient takes into account the heat transfer by convection, radiation, and between the particulates itself in the vibrofluidized layer; the heat losses into the surrounding medium are neglected.

It is well known that the drying process consists of two periods, namely the first period or period of constant drying rate and the second period of the fallen drying rate [2]. The temperature of drying material in the first period is constant and easy to determine (see Eq.3). The second period starts when the moisture content of solids reach the critical value, and the temperature of the solids are increasing. The temperature of the solids in this period is of the greatest interest.

The kinetics of the drying process in the second period is described by Luikov [3] as

$$-\frac{dX}{d\tau} = \chi N(X - X_{cr}) \quad (1)$$

where the experimental coefficient N depends on the drying parameters, and the relative drying factor χ depends on the properties of the material and on its initial moisture content.

There are two principal modes of operation, whether continuous or batch wise as it is schematically depicted in Fig.1. The batch wise operation mode is shown in Figure 1(a) and Figure 1(b) depicts the continuous operating mode. Items 1, 2, and 3 show vibrating container, material to be dried, and heating surface, respectively with I-entrance and II-exit of the material.

The equation of the heat balance for the batch wise mode of operation (Figure 1a) is similar to the one developed by Sheyman [4]. A similar balancing equation also describes the thermal treatment process for the continuous mode of operation (Figure 1b) with the following differences: there is a path differential $dx=vdt$ instead of the time differential dt and the specific heating surface area F_{sp} (the heating surface per unit of the bulk material displacement path) instead of the heating surface area F; and also the mass of the bulk of material in the container G is substituted by the mass of flow rate G^* in terms of the dry material.

The solution of the problem at hand is to determine the temperature of the drying material together with appropriate conditions which results as

$$\upsilon = \upsilon_1\left(\frac{y}{y_1}\right)^a - \frac{At_f}{\chi Na}\left[\left(\frac{y}{y_1}\right)^a - 1\right] - \frac{A(t_o - t_f) - B}{\chi N(a-1)}\left(\frac{y^a}{y_1^{a-1}} - 1\right) \quad (2)$$

where the following notations are accepted:

$$A = \frac{h_{ef} F}{G C_{ps}},$$

$$B = \frac{r \chi N \left(X_{cr} - X_{eq} \right)}{C_{ps}},$$

$$a = \frac{A}{\chi N \left(\frac{C_{ps} + C_{pl} X_{eq}}{C_{ps}} \right)},$$

$$y = \exp\left(-\chi N \tau\right), \quad \text{and} \quad y_1 = \exp\left(-\chi N \tau_1\right)$$

with t_1 as the time at the end of first drying period. The equation (2) expresses the relationship between the temperature of the solids and time.

From this analysis, with the surface temperature (*t*) being constant, the following relation for the first drying period (i.e. d?/dt = 0) is easily obtained as

$$\vartheta = t - \frac{rN}{C_{ps}} \quad (3)$$

This relation serves as a basis of the results compared later in the paper with experimental work.

EXPERIMENTAL RESEARCH

The experimental research was carried out in a setting as shown in Figure 2. It shows the main layout with conductive heat transfer. The following list identifies important components: (1) Container, (2) Vibrating table, (3) Frame, (4) DC electric motor, (5) Potentiometer, (6) Thermal probe, (7) Electro-tachometer transducer, (8) Secondary device of the electro-tachometer, (9) AC to DC converter, (10) Eccentric with variable eccentricity, (11) Electric heater, (12) Potentiometer, (13) Autotransformer, and (14) Chrome-copper thermocouple.

The moist material is loaded into the container (1) that has the electric heater (11) built into the bottom of the container. The level of electrical power is regulated by the laboratory autotransformer (13). Table 2 supporting the container is vibrating, and frequency of vibration, ranging from 0 to 50 Hz, is regulated by the DC electric motor (4). The speed of the electric motor rotation is fixed by the electric tachometers (7,8). The amplitude of the vibration is varied step-by-step from 0 to 5.0 mm by employing an eccentric mechanism with variable eccentricity. The experiments were carried out with constant and variable heating surface temperatures. The temperature of heating surface was regulated and measured by potentiometers (12,5) and by using the chrome-copper thermocouples mounted on the surface.

Temperatures of the solid particulates over the height of the layer were measured by a thermal probe. It consisted of 10 chrome-copper thermocouples, connected to the potentiometer (5). The parameters of the vibration in the experiments varied in the following range: the amplitude was within the limits A* =1-5 mm., and the frequency was f =10-50 Hz. The temperature of the heating surface was 150 - 300C, and the height of the dense particulates layer was h = 20 - 90mm. The rectangular container had a cross section of 55mm x 240mm, and its height was 350mm.

The material, crystal sodium chloride, with initial moisture content was loaded into the container which vibrated at a given amplitude and frequency, and it was dried to a constant weight.

Some experimental drying curves and the temperature variation of the material with respect to time are plotted in Figure 3. It contains t =165 ^{0}C, h = 44 mm, and f = 28.3 Hz. The curves 1, 1', and 1" are for A* = 4.5 mm; curves 2, 2', and 2" are for A* = 3.5 mm.; 3, 3', and 3" correspond to A* = 2.5 mm; and 4, 4', and 4" pertain to A* = 1.5 mm.

A significant portion of the drying process takes place at a constant drying speed in the firs period. During this period, the main quantity of moisture is removed at all drying conditions. The remaining moisture is removed in the second period, but the drying time is approximately equal in both the periods. It was also established that the drying was intensified with increase of the amplitude of vibration. Another parameter of vibration, frequency influenced the drying rate to a lesser extent. For example, the decrease of amplitude from A* = 4.5 mm to A* = 1.5 mm at constant frequency of f = 21.4 Hz resulted in decreasing the drying speed from N = 7.4•10^{-3} kg/kg/min to N = 5.1•10^{-3} kg/kg/min in the first period. That was approximately in 1.5 times. When the amplitude was constant at A* = 2.5 mm and the frequency decreased from f = 40 Hz to f = 14.7 Hz, the drying rate varied from N = 7.6•10^{-3} to 6.7•10^{-3} kg/kg/min. That rate was 1.1 times. These experiments were carried out at an average temperature of the heating surface temperature, t = 165 C, with the height of the initial layer of h = 44 mm and X_0 = 5.5•10^{-3} kg/kg.

The solid lines in Figure 3 are the temperature curves calculated according to the equations (2) and (3). In all these cases, the curves with calculated values are somewhat higher than those with the experimental measurements. The deviations do not exceed 10%.

CONCLUSIONS

In this paper, a specific case of the thermal treatment of the fine particulates in vibrofluidized layer is considered. Both theoretical and experimental research results are presented for comparison. Differential equations (2) and (3) are formulated for both batch wise and continuous modes of operations covering all possible cases. The solutions of the formulated equations for the variable and constant temperatures of the heating surface are obtained to reach the stated objectives. These solutions resulted in establishing the relationship between temperature of solids to be dried and time of the process. That in turn allows to control the quality of the drying materials. A comparison of the experimental data of the temperatures of solids with those of the theoretical equation shows a very good result with maximum deviation not exceeding 10%. The proposed equations may be used for a variety of materials and with appropriate conditions for a number of thermal treatment processes.

ACKNOWLEDGEMENTS

The authors would like to thank the ASME Mechanical Engineering Technology Department Heads Committee to organize these paper sessions for the Engineering Technology community. This action has provided an important venue for professional development of faculty. Also, the Wayne State University Division of Engineering Technology deserves a note of thanks for continuing financial support for travel to conferences and giving encouragement for scholarly participation.

REFERENCES

1. Syroedov, V.I., Heat Transfer in Vibrofluidized Bed at Constant Temperature of the Heating Surface, Heat/Mass Transfer MIF – 92, Belarus, Minsk, 1988.
2. Keey, R.B., Introduction to Industrial Drying Operations, Pergamon Press, 1978.
3. Luikov, A.V., Heat and Mass Transfer in Capillary-Porous Bodies, Pergamon Press, London, 1977.
4. Sheyman, V., In "Heat and Mass Transfer", **6**, part 1. Science Thought, Kiev, 1988.
5. Holman, J.P., Heat Transfer, 8-th Edition, McGraw-Hill, Inc.

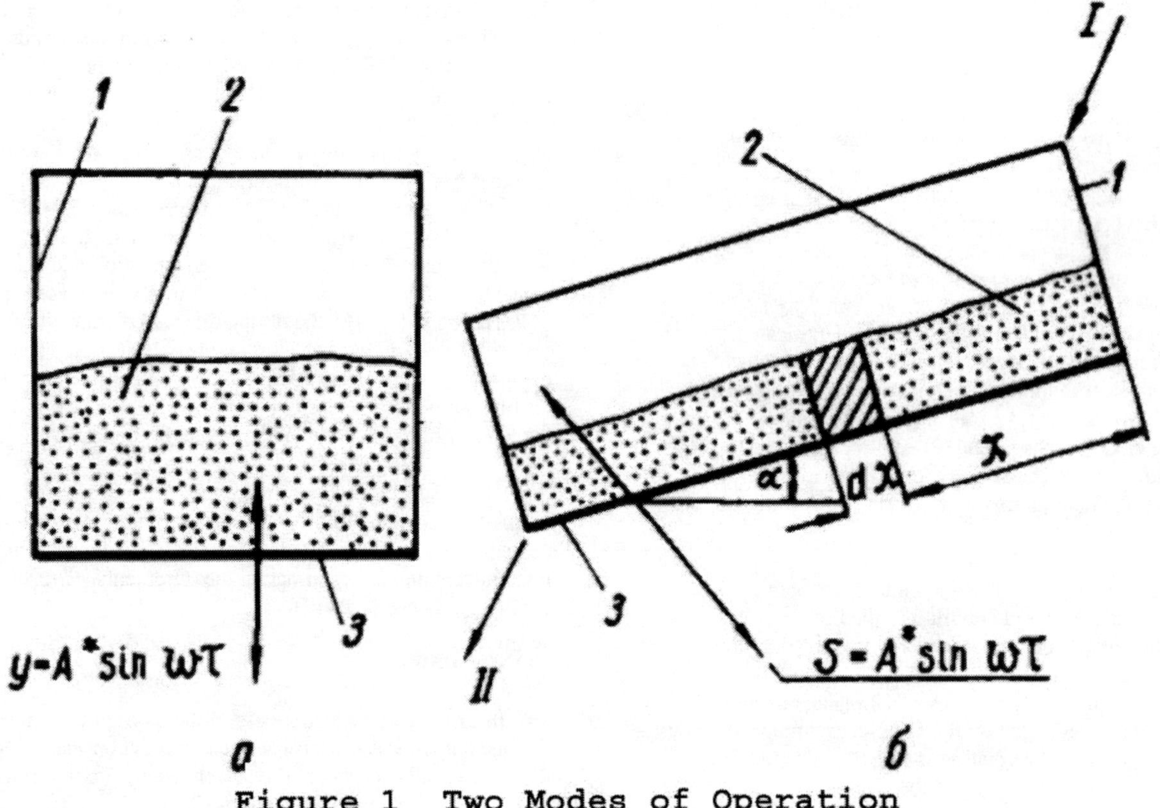

Figure 1 Two Modes of Operation

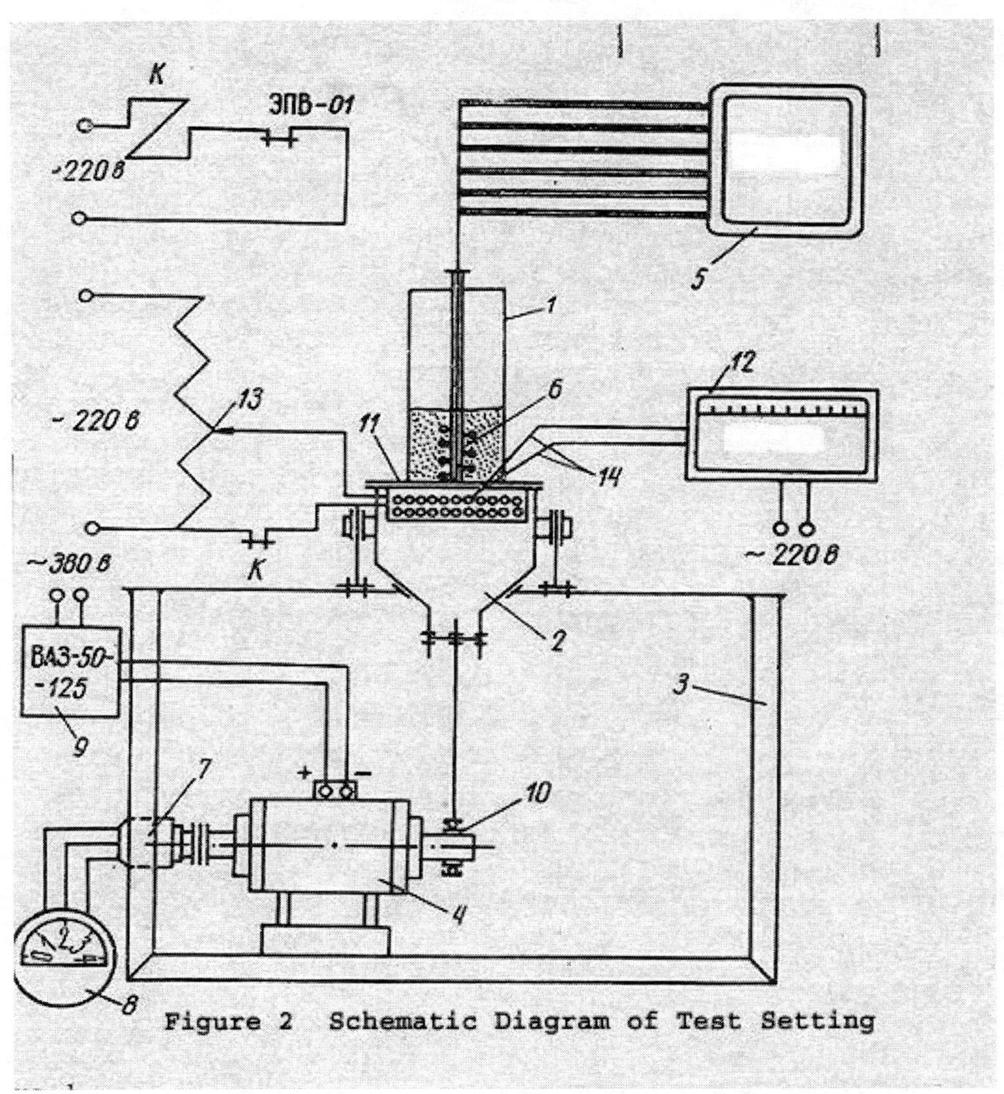

Figure 2 Schematic Diagram of Test Setting

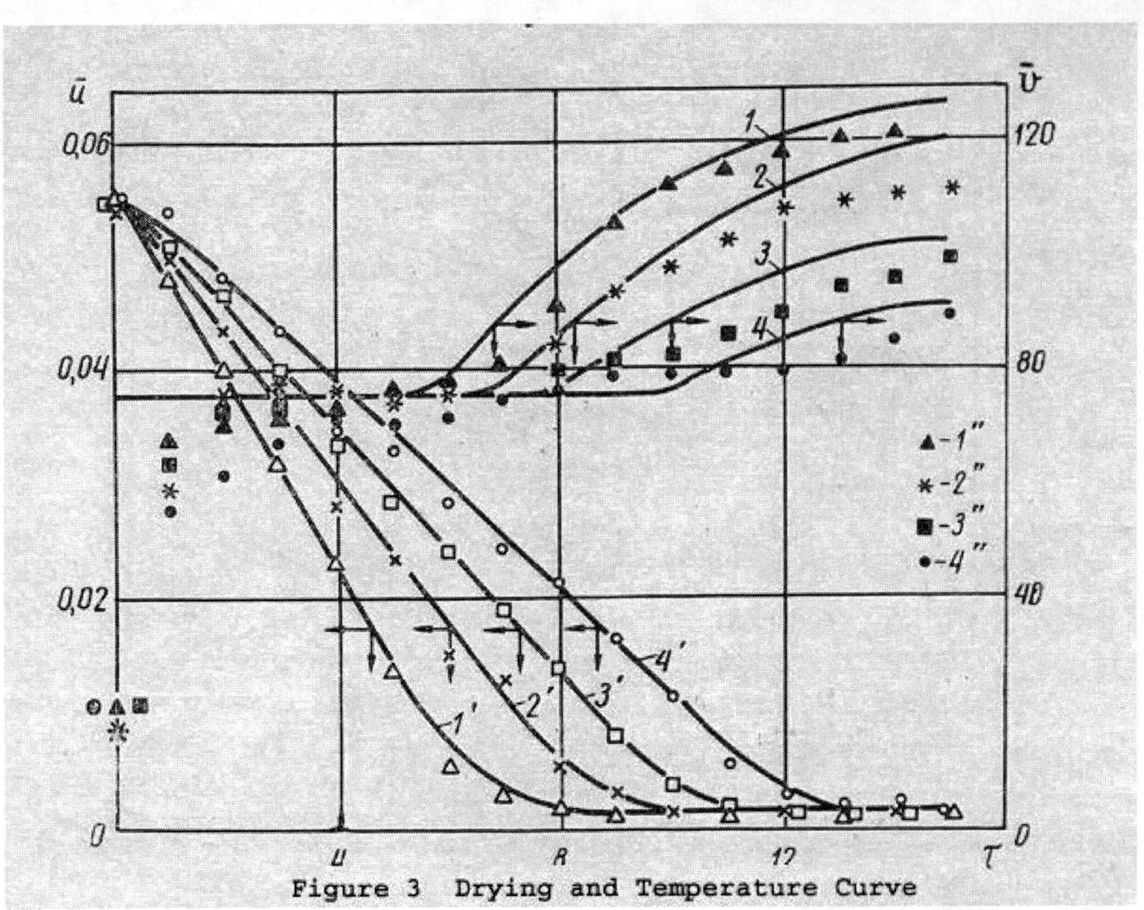

Figure 3 Drying and Temperature Curve

Proceedings of IMECE'03
2003 ASME International Mechanical Engineering Congress
Washington, D.C., November 15–21, 2003

IMECE2003-41974

METHODOLOGY FOR SELECTION OF RAPID TOOLING PROCESS FOR MANUFACTURING APPLICATION

Chan W. Chung[†], Joon Hong[†], Alex Lee[†], Karthik Ramani[†], Mileta Tomovic[††]
Purdue Research and Education Center for Information Systems in Engineering (PRECISE)
Purdue University, West Lafayette, Indiana
[†] School of Mechanical Engineering, [††] Mechanical Engineering Technology

ABSTRACT

Digital representation of geometry is transforming the methods for the manufacturing of products. Such transformations include a significant reduction in cost and lead-time to produce products and to increase the variety of products. The ability to produce tooling from material removal and additive processes has led to new methods to produce tooling from digital geometry representations. These include processes that are grouped together as fast-free-form-fabrication (FFFF) methods. In addition, other possibilities exist in modular tooling for generating variety and new processes/methods for producing such geometries. Also, conventional machining processes have improved significantly. As a result of these transformations, a tooling designer faces many choices of methods, materials, and lead-times to produce tooling. These decisions involve geometric reasoning coupled with other requirements based on production needs, cost constraints, and individual casting shop capabilities.

This paper discusses a methodology for selection of rapid tooling (RT) processes based on a number of user-specified attributes and relative cost and lead-time comparisons across a wide spectrum of available RT processes. The method is based on a combination of weighted functionals and on the approximate mathematical models of metalcasting process costs and their lead-time. The method has been put on the Internet for easy access. It is currently limited to only several of the most common RT processes and materials. However, the database will be expanded in the future to include a majority if not all of the metalcasting processes.

INTRODUCTION

With today's rapidly changing technology and economy, introducing a product to the market quickly is becoming more important than ever.

Application of different rapid tooling processes can significantly reduce the manufacturing time and cost, especially in short run production. Rapid Tooling is a general term that is used to describe different methods of fabricating production tools (molds) rapidly. Many of the current rapid tooling methods in injection molding, sand casting, and investment casting make use of rapid prototyping.

Rapid Prototyping (RP) was introduced commercially in the 80s, but a study shows that only 15% of the mechanical engineering sectors are using the technology [1].

According to the study, rapid prototyping technologies are not used to their full potential because many engineers are not aware of their capabilities. About 28% of large companies (those with more than 100 employees) have used Rapid Prototyping technology, but only 8% of those with less than 100 employees use the technology in the mechanical engineering area [1]. In general, metalcasting companies are small to medium companies. Cadalyst points out that most of the 3,000 foundries in the United States do not take advantage of rapid prototyping [2].

Many researchers have explored and developed different techniques to make patterns and tooling for foundry using rapid prototyping technology [3-9]. Related works include a build time comparison study using different RP methods [10]. Paxton benchmarked different rapid prototyping processes [11]. Wang developed an integrated decision process using RP selection criteria and decision factors [9]. However, his work is limited to tooling master fabrication and does not address the direct tooling process.

More detailed analysis of inter-related parameters of different manufacturing processes must be done in order to build a foundation database for a tooling advisory system. The following text presents a review of the characteristics of different RP processes and discusses the RP selection criteria. The overall structure of the RP Advisory system and its implementation is also discussed

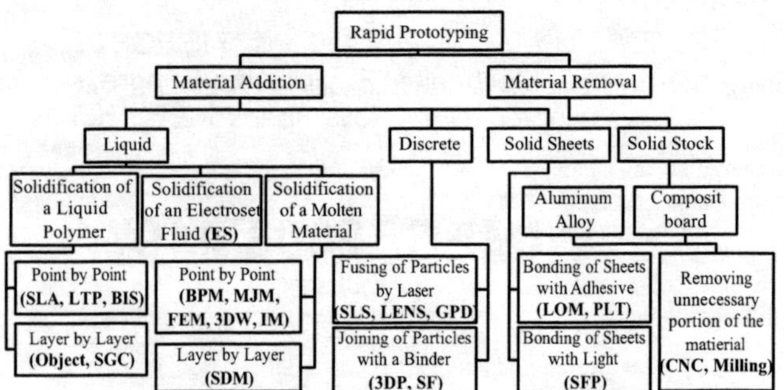

Figure 1: Classification of rapid prototyping method [4]

TYPES OF RAPID PROTOTYPING METHODS

There are many different rapid prototyping methods. Kenneth [4] has proposed a classification system for rapid prototyping methods based on the way parts are built. Figure 1 illustrates his proposed classification system (modified to include the material removal process). Most of the rapid prototyping processes are classified in the category of "Material addition process." The most commonly used RP processes for tooling manufacturing are Stereolithography (SLA), Fused Deposition Modeling (FDM), Selective Laser Sintering (SLS), Laminated Object Modeling (LOM), and 3D Printing (3DP) [3]. Based on the authors' current interaction with companies [25, 26], it was found that high speed CNC machining of different pattern materials (composite board, Al alloys) can be categorized as an RP method.

Stereolithography (SLA)

The SLA (Stereo Lithography) method uses ultraviolet (UV) light, e.g. Helium-Cadmium or Argon ion laser. When photosensitive liquid resin is exposed to ultraviolet light, it forms a solid polymer. A laser scans the model shape of each layer. After the laser is applied to the layer, an SLA machine lowers the platform, and the blade wipes the resin to make the uniform layer-thickness above the part. The process is repeated until the whole part is built. SLA is generally highly accurate and is capable of producing a model with a good surface finish, complex geometry, and thin-walls with high precision requirements [9].

Laminated Object Modeling (LOM)

Another method that is used often is Laminated Object Modeling (LOM). LOM uses a build material that contains a heat-sensitive adhesive. The build material is bonded to the previous layer using a hot roller. The heat activates the adhesive and helps the current layer bond to the previous layer. The laser is then applied to trim and hatch the exterior area. A laser power is precisely controlled to penetrate only to a single layer. The process is repeated until the entire part is built. The exterior acts as a support during the process and is removed later. A common material for this process is a thin sheet of paper. The finished model, when made with paper, has the surface finish and consistency of wood and is suitable for large, complex, and bulky parts. Post processing is done manually and is time-consuming. This process is becoming less popular than other processes due to the poor surface quality and precision.

Fused Deposition Modeling (FDM)

The Fused Deposition Modeling (FDM) machine uses a thread of molten polymer to build the model. The build material is heated above its melting point and extruded from the nozzle, bonds to the previous layers, and solidifies. Once a layer is completed, a new layer can be applied until the build process is done. It can build a model fast with high accuracy and good surface finish. The resulting part will have good strength, and in many cases, support structures are needed in this process [12].

Selective Laser Sintering (SLS)

The Selective Laser Sintering (SLS) process uses a fine powder and a CO_2 laser. A laser is applied to fuse the powder. Before the powder is sintered, the whole bed is pre-heated just below the melting point to minimize the effect of distortion and to facilitate fusion to the previous layer. After the laser is drawn in the layer, the bed is lowered and powder is spread evenly over the build area using a counter-rotating roller. Un-sintered powder acts as a support during the process and it can be recycled after the build is completed. This process does not need a support structure and can produce very strong metal composite parts.

Other than the methods listed above, there are many more, such as 3DP (3D Printing), and MJM (Multi-Jet Modeling). Table 1 lists the basic characteristics of each representative method in RP.

Table 1: Basic specification and application of RP methods

Method	Building Speed[1]	Size limits (mm)	Materials	Major application
SLA	Slow - Fast	250x250x250 – 508x508x600	Photo-polymer (epoxy)	Prototyping, Investment Casting Pattern
FDM	Medium	254x254x254 – 600x500x600	ABS, Wax	Prototype, Investment Casting Pattern
SLS	Medium - Fast	250x250x200 – 700x380x580	Any powder, Nylon, Metal powder, Polycarbonate, Polystyrene	Pattern for Investment Casting, Tool for Injection Molding
3DP	Medium - Fast	203x254x203 – 500x600x400	Starch-based powder, Plaster-based powder	Modeling, Direct Shell Fabrication for Casting
LOM	Slow – Medium	381x254x356 – 813x559x508	Paper, Composite	Foundry Pattern, Rapid Modeling

TOOLING ADVISORY SYSTEM

The overall metalcasting process diagram is shown in Figure 2. The dotted square marked "A" represents the Tooling Advisory System within the metalcasting process. As shown in the figure, the advisory system will be used in the early phase of the manufacturing process. Manufacturing decisions are typically made in each different step based on customer requirements. The RT (Rapid Tooling) advisor will help users to negotiate between multiple variables in order to meet the desired goal.

[1] The Building Speed of an RP Machine depends heavily on the setting of the machine.

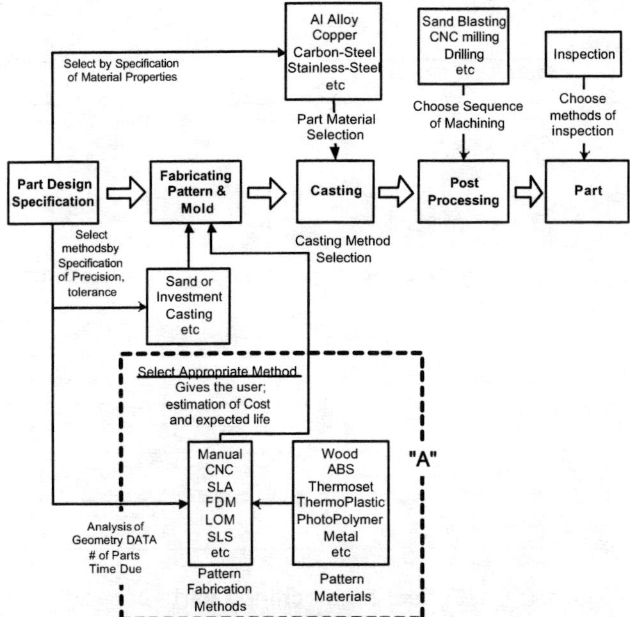

Figure 2: Structure of the Foundry process and the position of the Tooling Advisory System

There are various factors that determine the RP building speed and cost of tooling. The basic parameters that influence the speed of fabricating tooling are related to the performance of the machine itself as well as to the RP material For example, the power and speed of the laser are very important factors in SLA and SLS that affect build time. Given the performance of a machine, users can also control the machine parameters to trade off between the building speed and the quality of the parts. Although the machine performance can vary based on user settings, it was found that there is a general agreement on the average capability of different machines.

The factors that are examined in this paper include material properties, part geometry, RP time, and cost. The tooling advisory system has a database with detailed information on the common rapid prototyping systems, along with a set of rules to differentiate among various RP processes.

RP MATERIAL PROPERTIES

RP materials have their parameters typically optimized by the manufacturers for a specific RP machine. Even for the same machine, some materials take more time to cure than others, and some materials are better suited for use in building precision parts than others. Although the number of RP materials is limited, their properties can vary greatly. Some of the important material properties include: hardness, density, yield strength, tensile strength, elongation, flexural modulus, impact strength, and creep. The physical properties of the final RP part depend on how the model was built and the type of finishing process. For example, the properties of the materials used in 3DP (3D-Printing) vary significantly depending on whether or not any material infiltration was performed and on the type of infiltration material. Also, in the LOM process, the coating material on the surface of the finished part determines the durability and stability of the part. Different SLA machines use different lasers, and because the wavelengths of the lasers vary, different photopolymers have to be used in different machines. Table 2 shows the typical properties and ranges of a common RP material.

Table 2: Typical properties of RP materials (RenShape® SL5220) [13] [14]

Measurement	Method	Value
Appearance		Clear Amber
Density		1.15 g/cc
Hardness, Shore D	ASTM D2240	84 – 85
Flexural modulus	ASTM D 790	1,724 – 3,061MPa (250 - 444 KSI)
Flexural strength	ASTM D 790	44 - 74MPa (6,400 – 10,700 PSI)
Tensile modulus	ASTM D 638	1,455 – 3,020Mpa (211 – 438 KSI)
Tensile strength	ASTM D 638	15 – 45Mpa (2,200 – 6,500 PSI)
Elongation at break	ASTM D 638	1.2 – 1.6%
Impact strength, notched Izod	ASTM D 256	21 J/m (0.4 ft – lbs/in)
Thermal onductivity		0.141 W/m °K (3.37 x 10^{-4} cal/sec.cm.°C)
Density		1.18g/cm^3

In addition to the material properties listed above, there are other important properties that affect selection of the pattern material.

Durability is one of the most important properties that a designer considers when he/she selects a pattern material, since it is directly related to the number of parts that can be produced with the pattern. Durability can be considered as a combination of several different properties of a material. Although not accurately and analytically correlated, durability is related to strength-related material properties such as hardness and wear-resistance. Durability indicates how long a pattern (RP part) will last and is directly related to the number of molds and cores it can produce. Wear resistance is influenced by several material properties: $W_{ab}^{-1} = \rho LA / \Delta m$, where, ρ: density of the material, L: total length of the wear path, A: contact area, and Δm: mass loss. The wear-resistance is relatively proportional to the hardness of the material. The toughness (another factor influencing the pattern life) of a material depends on the bending strength and fracture strength of a material [16]. The pattern life expectancy also varies depending on several parameters other than the material itself. The speed of production, pressure, or operation temperature, for example, can also affect the pattern life. Table 3 shows a simple comparison of the life expectancy of different pattern materials. For example, an epoxy-based pattern used for the sand casting application can typically produce up to 5,000 molds [3]. The life expectancy of tooling made with RP materials can be compared with *Wood* or *Epoxy* tools, Table 3.

Table 3: Tooling wear inspection interval for different pattern material for Sand Casting

Material	Life Expectancy	Wear Check Interval
Wood	500~1,500	1,000
Epoxy	3,000~5,000	1,500
HMP	25,000~40,000	7,500
Aluminum	30,000~50,000	10,000
Brass	100,000~200,000	25,000
Gray Iron	200,000~300,000	25,000
Ductile Iron	300,000~500,000	25,000
Stainless	500,000~750,000	25,000

Table 4 shows some examples of tooling process, typical building time, cost, and life of the patterns [15].

Table 4: Comparison of various tooling techniques. Partly cited from [15]

Process	Time	Quantity	Cost
CNC machining	2-6 weeks	1K-10K	$2K-$20K
Aluminum tooling	10-12 weeks	50-20K	$10K-$20K
Silicon rubber	1-2 weeks	1-50	$1K-$5K
Epoxy composite	6-8 weeks	50-2K	$2K-$10K
Keltool	4-6 weeks	50-20K	$2K-$7K
Spray metal	3-6 weeks	50-2K	$2K-$10K

Chemical resistance is also an important property to consider when the part is exposed to a chemically harsh environment (oil, water, acid, hydroxide, etc). Thermo set plastics (Photopolymer, Epoxy) generally have very good chemical resistance. Epoxy based materials exhibit good resistance against most acids, oil, and hydroxide, but poor resistance to acetone or ketone-related chemicals. It is hard to categorize all the materials related to RP systems, but a general categorization can be done based on the typical foundry environment. In this application, more weight is applied to the sand casting processes.

The surface quality of a part is directly related to the grain size of the material. The surface quality of an RP part is related to the layer thickness, beam size, or nozzle size (when it is an FDM part). The surface quality depends on the machine capability and the type of RP material used. The accuracy of an RP part is determined by the layer thickness and beam diameter (size of nozzle). The surface roughness is related to the accuracy of the RP part. In practice, if an RP part is used as tooling, some additional finishing and polishing process is required in order to eliminate the stepped surface that is associated with most rapid prototyping processes. Therefore, the machinability of the cured RP material must also be considered.

Machinability is a term that is used very frequently by many engineers, but yet is very hard to define. Machinability can be represented in terms of the material removal rate, the length of tool life, or the edge definition of the machined parts [17, 18]. In the context of this work, machinability means the ease in finishing or machining the RP part. For example, the paper-based model made by the LOM process is soft but is very difficult to machine and requires a manual sanding operation. Alternately, SLA parts are easy to machine even though the material itself is relatively hard. These material properties are stored in a database and are used in analyzing and recommending the appropriate RP tooling method.

PART GEOMETRY

In many cases, the cost of metalcastings is driven by their size, while geometric complexity is considered as the secondary cost-driving factor. Unfortunately, there is no accurate way to directly determine the influence of part complexity to the cost [19]. However, statistical approaches relating the geometric complexity to cost have been done in some research [20-22]. Bribiesca proposed a method of determining relative complexity using surface area to volume ratio [19]. Swift developed a systematic method to categorize a part based on its complexity. Figure 3 shows an example of this categorization system on prismatic parts.

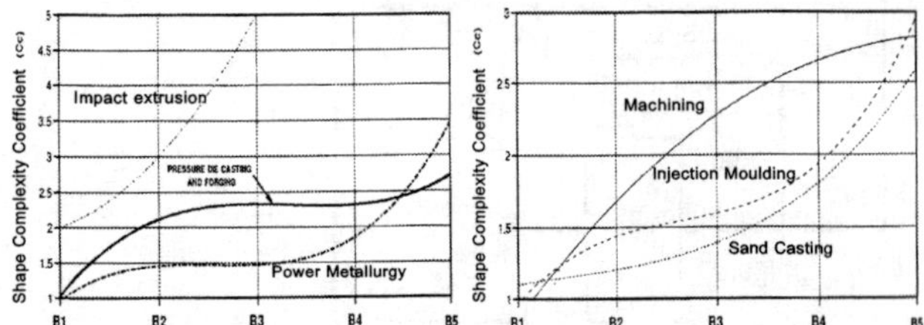

Figure 3: Categorization of prismatic parts based on complexity [20]

Figure 4: Cost effect on machined parts based on shape complexity [20]

Swift determined a correlation between the complexity based on the categorization and cost of machining (Figure 4). By using the shape complexity function, one can estimate the cost index for different parts.

The Designer Influence Cost Element (DICE) factor also provides a systematic way of estimating the cost of metalcastings [22]. The DICE method gives a reasonably accurate cost estimate for a large production quantity, but not for low production volume. Another drawback of the DICE method is that it requires in-depth knowledge about the part and the casting processes. Boothroyd [21] tried to estimate cost by looking at the number of patches for a given part. This method is applicable only for conventional machining and cannot be applied to the RP process. In general, these methods are not applicable to rapid prototyping because the rapid prototyping process is a material addition process rather than a removal process.

BUILD TIME OF RP PARTS

Height, width, and depth of a part represent the part's "Box Volume." The "Box Volume" determines the overall size of a part and whether or not it can be made as a single piece with a specific RP machine. The workspace of an RP machine should be approximately 20% bigger than the box volume of the part or the segment of a desired part. When a desired part is bigger than the workspace of the machine, it is divided into segments and assembled later. The most significant factor affecting build time is the height (z-direction) of a part since the layer transition interval often takes more time than the scanning (curing photo-polymer material by traveling laser beam on the surface) of a single layer. The cost of an RP part is also directly proportional to its volume, and the cost of the material is typically 20-30% of the overall cost [23]. In general, complexity is not a dominating factor for RP machines. Most RP processes can produce a complex part without any significant difficulty. Therefore, in most cases, the time it takes to build an RP part is directly proportional to the Box Volume.

The only exception is the LOM process. Since LOM cuts the perimeters of each layer, the complexity of the part affects the building time significantly. The relative cost and time estimation functions are determined and presented in the later sections of this work.

Build Time and Cost Estimation of SLA, FDM, and SLS

The building time and cost of RP parts are related to the machine specifications and operating parameters. Table 5 lists the different parameters that are used to estimate the building time and the cost. The average process parameters, e.g. scanning time and time between building layers, are determined and stored in the Rapid Prototyping database. Modifying one parameter often causes other parameters to change as well. For example in the SLA process, an increase in the layer thickness increases the laser density or decreases the scanning speed.

Table 5. Parameters used to estimate building time and cost

d_b: Beam Diameter or Nozzle diameter (mm or inches)
d_d: Part Depth (mm or inches)
d_h: Part Height (mm or inches)
d_w: Part Width (mm or inches)
d_l: Layer Thickness (mm or inches)
a_s: Surface Area of a part (mm^2 or square inches)
V: Part Volume (mm^3 or cubic inches)
r_s: Scanning Speed (mm or inches/seconds)
t_b: Building Time (seconds)
t_c: Curing Time (seconds) - *Proportional to the Volume of the RP Part*
t_f: Finishing Time (seconds) - *Proportional to the surface area of the RP Part*
t_i: Time Interval between Layers (seconds)
t_o: Overall Time (seconds)
t_p: Preparation Time (seconds) *Preparation of the CAD file & setting up the machine*
c_h: Hourly Rate (dollars/hour) (Labor, price of the machine, and cost of Electricity /hr)
c_b: Building Cost (dollars)
c_m: Material Cost per unit volume (dollars/ mm^3 or cubic inches)
c_o: Overall Cost (dollars)
c_{oh}: Over-Head (dollars)
c_e: Electricity cost (dollars/Watts)
p_e: Electricity used (Watt/hour)
Hg*: Hatching Grid length density (mm or inch / mm^2 or square inch)
* The ratio of the length of hatching grid over the area. If the length of the grid is 2 inch on the area of 1 square inch then the factor becomes 2. The hatching grid is necessary for de-cubing (taking the unnecessary part out from the bulk shaped material stack) purpose.

The time it takes to build a part using RP can be estimated by adding the time to build parts and the transient time (going from one layer to the next). This is expressed with Equation 1.1

- **Net Building Time:** $t_b = \left(\frac{V}{d_l d_b r_s}\right) + \left(t_i \frac{d_h}{d_l}\right)$ (1.1)

Once the net building time is known, the overall time can be estimated by adding preparation time, building time, curing time, and finishing time (Equation 1.2).

- **Overall Time:** $t_o = t_p + t_b + t_c + t_f$ (1.2)

By multiplying the corresponding cost rate for different components such as electricity consumption and machine operating cost, part build cost can be obtained (Equation 1.3).

- **Building Cost:** $c_b =$

$$c_m V + \frac{1}{3600} t_b (c_h + c_e p_e) + \frac{1}{3600} t_c (c_h + c_e p_e) \quad (1.3)$$

The overall cost can be calculated by adding preparation cost, finishing cost, building cost, and overhead cost (Equation 1.4).

- **Overall Cost:** $c_o = \frac{c_h}{3600}(t_p + t_f + t_c) + c_b + c_{oh}$ (1.4)

Equations 1.1 through 1.4 are applicable to SLA, FDM, and SLS processes. The building times for these machines are roughly proportional to the Net Volume of the parts when the height of the box volume is the same. From Equation 1.1, it is clear that the building time t_b has two components. One is the time it takes the laser to build the part by scanning the cross-section, and the other is the time for changing the layer. These results can vary greatly from machine to machine depending on the manufacturer's specifications and the condition of the machine.

Building Time and Cost Estimation of LOM

For the LOM process, the surface area or the perimeter of each slice determines the time for building the parts. There are three components in the estimated cost:

- ❑ Cutting the perimeters of each layer to build net shapes
- ❑ Layer transition and pressing to bond each layer
- ❑ Cutting boundary and grids of each layer

The time that it takes to build a part in LOM can be estimated by adding the time that it takes to cut a part shape, the transient time from one layer to the next, the edge cutting time, and the hatching grid cutting time (Equation 2.1).

- **Net Building Time:** $t_b =$

$$\frac{a_s}{\sqrt{2 d_l r_s}} + \left[t_i + \frac{2(d_w + d_d)}{r_s}\right] \times \frac{d_h}{d_l} + \frac{H_g (d_w d_d d_h - V)}{r_s \times d_l} \quad (2.1)$$

Then by adding the preparation time, the building time, and the finishing time, the overall time can be calculated (Equation 2.2)

- **Overall Time:** $t_o = t_p + t_b + t_f$ (2.2)

The building cost and overall cost can be obtained by multiplying the time and associated cost rate (Equation 2.3 and 2.4).

- **Building Cost:** $c_b = c_m V + \frac{1}{3600} t_b (p_e c_e + c_h)$ (2.3)
- **Overall Cost:** $c_o = \frac{c_h}{3600}(t_p + t_f) + c_b + c_{oh}$ (2.4)

Equations 1.1 through 2.4 give an estimation of the building time and the cost of making the parts with the rapid prototyping process. These equations can be used as a relative index to compare different processes.

SYSTEM ARCHITECTURE

The tooling advisory system is constructed using JSP (Java Server Page) and JAVA programming language in conjunction with Oracle® DBMS (Database Management System). The system is implemented on a Tomcat® JSP server running on a Windows 2000® machine. The JAVA programming language is ideal for writing object-oriented programs, and Oracle® DBMS is well known for its stability and scalability. The structure of the overall system is shown in Figure 5.

The system stores user inputs (geometry information, material properties, etc.) in the form of JAVA Beans (as internal variables stored in the server memory). The system then retrieves RP material from the existing material database that meets the user specifications. And the system performs various cost estimations and reasoning based on the process information. The current system has the following characteristics:

- ❑ Web-based application - Dynamic web pages
- ❑ Server – Client system model
- ❑ Relational database management
- ❑ 3-Tier architecture (User view layer - Application layer - Database layer)

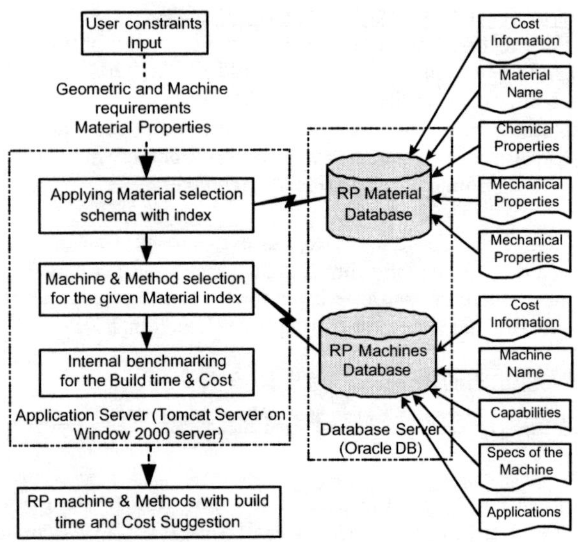

Figure 5: Structural diagram of the Advisory System

The material attribute table contains all the information about the material properties including the unit cost. In addition, there can be different machine-material combinations. Thus, in order to eliminate redundancy in the database tables, Jeffrey recommends using the association table [24]. The data structure of this system applied to this problem is illustrated in Figure 6.

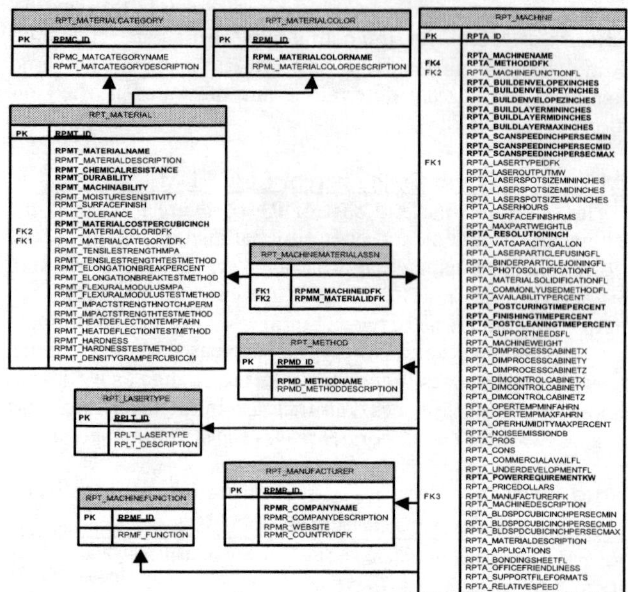

Figure 6: Database structure of the Advisory System

GRAPHIC USER INTERFACE AND TESTING

During the early stages of design, the designers consider the material within the general scope of its application. Requesting and specifying the exact values of different physical properties is not very practical in the early design stages. Therefore, providing intuitive estimation inputs is in many ways more appropriate. Figure 7 shows the screenshot of the advisory system input menu. The parameters that are considered include geometric information (part volume, surface area, width, and height), and desired part properties (material type, durability, chemical resistance, machinability, and surface quality).

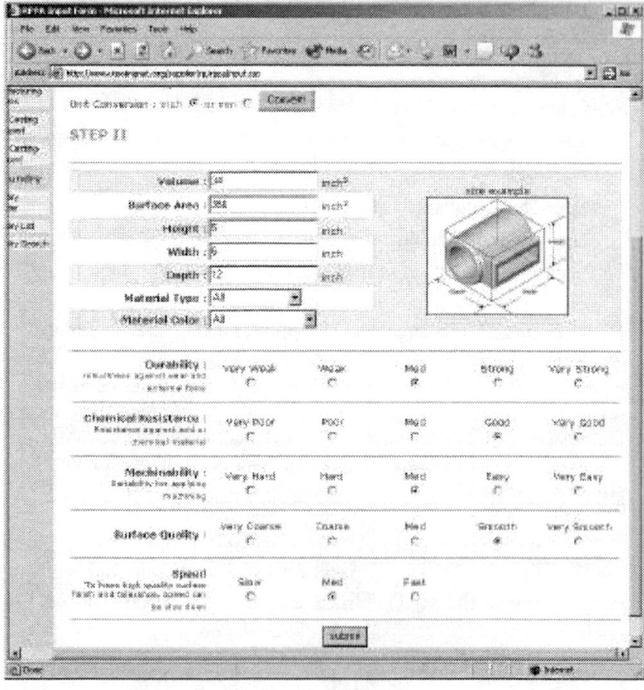

Figure 7: User input screen of the Advisory System

Upon entering the requested information, the system begins searching for suitable materials and machines. And it calculates the relative times and costs required to build an RP part based on the information in the database and the logic built in the application.

Table 6: Applicability between casting processes and pattern materials

	Typical Pattern Material	SLA	FDM	3DP	SLS	LOM	Comp. Board	Metal
Sand	Wood, Metal	1	1	0	1/0	1	1	1
Shell (Heating)	Metal	0	0	0	0	0	0	1
Investment (Direct)	Wax	1	1	1	0	1/0	0	0
Investment (Indirect)	Metal, Al	1/0	1/0	0	1/0	0	1/0	1
Die	Metal	0	0	0	0	0	0	1
Centrifugal	Metal, Ceramics	1/0	1/0	1/0	1/0	1/0	1/0	1
Ceramic Mold	Metal	1/0	1/0	0	1/0	0	1/0	1
Plaster Mold	Metal	1/0	1/0	0	1/0	0	1/0	1
Lost Foam (Indirect)	Metal, Al	1/0	0	0	1/0	0	0	1

1: Applicable
1/0: Applicable conditionally
0: Non-applicable

Figure 8 is the output screen. It shows the machines that can be used, and the cost of the materials, time estimation, and descriptions of each RP process. When the size of a part to be made is bigger than the machine envelope, the system advises the user to divide the part into smaller segments. The system can also sort the results based on user preferences.

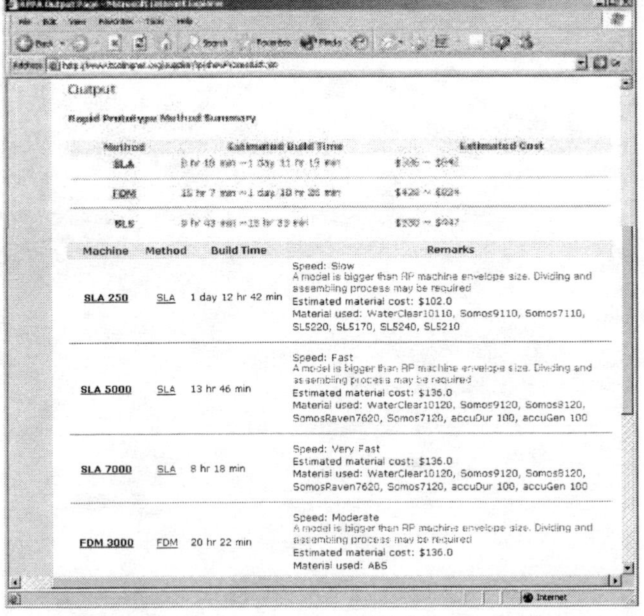

Figure 8: Output screen of Advisory System

CONCLUSION

A prototype of the Rapid Tooling Advisory System has been developed in order to assist engineers in the selection of a proper method under specified requirements. The focus of the project is on the rapid tooling methods that can be realized by using rapid prototyping processes, especially for the metal-casting industry. A required database of RP processes and materials has been developed and populated. The relationships among the data have been identified and have been incorporated into the database in the form of relation tables. As a result, a web-based application was successfully created.

The functional relationship between the input and output has been tested and has resulted in intuitive feedback to the engineers. Future work will focus on expansion of the current system and further testing of the system in industry.

ACKNOWLEDGMENTS

The authors wish to express their gratitude to their colleagues from the Advanced Technology Institute, Charleston, South Carolina and to the American Foundry Society, Des Plaines, Illinois, for their collaboration and input on this paper. The authors also wish to thank the Defense Logistics Agency for funding this project under U.S. Government Contract No. SPO103-00-C-0004.

REFERENCES

1. CAD digest "User Views on Rapid Prototyping," http://www.caddigest.com/subjects/research/select/cadspaghetti_rp.htm
2. Cadalyst http://www.cadalyst.com/features/0501rapid/0501rapid.htm
3. Pham, D.T. "Rapid manufacturing: the technologies and applications of rapid prototyping and rapid tooling," Springer-Verlag London Limited (2001)
4. Kenneth G. Cooper "Rapid Prototyping Technology," Marcel Dekker, Inc (2001)
5. Detlef Kochan, Chua Chee Kai, Au Zhaohui "Rapid prototyping issues in the 21st century," Computers in Industry 39 (1999) 3-10
6. Wohlers, Homepage: http://www.wohlersassociates.com/
7. Bernhard Mueller, Detlef Kochan "Laminated object manufacturing for rapid tooling and patternmaking in foundry industry," Computers in Industry 39 (1999) 47-53
8. R. Gustafson, E. Guinn, and D. Tait, "Rapid prototyping for pattern and foundry tooling," Modern Casting. v. 85 n2 Feb 1995. p 48-50. (1995)
9. W. Wang, J. Conley, H. Stoll, R. Jiang, "RP process selection for Rapid Tooling in sand casting," Proceedings of the 10th Solid Freeform Symposium, University of Texas at Austin, August 9-11, (1999)
10. G. Mensing, I. Gibson, "Building time estimations for large scale modeling," Proceedings of the 9th solid Freeform Fabrication Symposium, University of Texas at Austin, August 10-12 (1998) 343-350
11. J. Paxton, "Benchmarking rapid tooling process," Rapid prototyping and manufacturing 1999 – advanced product development Technologies in Action, v. 3, April 20-22, 1999, Rosement Convention Cener, Rosement, Illinois. 519-603
12. Stratasys, Inc. Homepage, http://www.stratasys.com
13. ATI rapid, Homepage, http://www.atirapid.com
14. Vantico, Homepage, http://www.vantico.com
15. G. Jack, Z. Zhou, "Rapid pattern based powder sintering technique and related shrinkage control," Materials and Design 19 (1998) 241-248

16. Hans Berns, “Comparison of wear resistant MMC and white cast iron,” Wear 254 (2003) 47-54
17. Ductile.org, Homepage, http://www.ductile.org/didata/Section6/6intro.htm
18. Brass.org, Homepage, http://www.brass.org/Training/Lecture
19. E. Bribiesca “A Measure of Compactness for 3D Shapes,” Computers and Mathematics with Applications 40 (2000) 1275-1284
20. KG Swift and JD Booker, “Process selection from design to manufacture,” Arnold (1998)
21. Geoffrey Boothroyd, Peter Dewhurst, Winston Knight, “Product design for manufacture and assembly,” Marcel Dekker, Second Edition Revised and Expanded (2002)
22. Bryan R. Noton, "ICAM MANUFACTURING COST/DESIGN GUIDE FINAL TECHNICAL REPORT VOLUME II - AIRFRAMES USER'S MANUAL," AFWAL-TR-83-4033 (1983)
23. T. Wohlers, T. Grimm, “The Real cost of RP what to consider before bringing rapid prototyping in-house,” Time-Compression Technologies, March/April (2002)
24. Jeffrey A. Hoffer, Mary B. Prescott, Fred R. McFadden “*Modern database management*” Prentice Hall, sixth edition (2001)
25. Cunningham Pattern & Engineering, Inc. 4399 US 31 North Columbus, IN Homepage, http://www.cunn.com
26. Clinkenbeard & Associates, 577 Grable St. Rockford, IL Homepage, http://www.clinkenbeard.com

Proceedings of IMECE'03
2003 ASME International Mechanical Engineering Congress
Washington, D.C., November 15–21, 2003

IMECE2003-42158

PARAMETRIC STUDY OF SURFACE FINISH IN END MILLING USING ROBUST DESIGN TECHNIQUES

Alok K. Verma/Old Dominion University **Steven L. Holcomb/Northrop Grumman Newport News**

Paul Blessner/Northrop Grumman Newport News **Dave Tilman/Northrop Grumman Newport News** **William F. Johnston/Northrop Grumman Newport News**

ABSTRACT

Surface finish in end milling depends upon a number of variables, such as feed rate, cutting speed, tool material, etc. The relative effect of these variables on surface roughness is not understood very well. The Taguchi method is used in end milling process to identify variables having major influence on surface finish. Partial factorial design using L_9 orthogonal array is used in the design of experiments. Signal to Noise ratio analysis along with analysis of variance is used to study the effect of these parameters on surface finish. Knowledge about these relationships will help a process planner or a machinist optimize the cutting process with respect to surface finish.

Key Words: Parametric Study, Milling, DOE, Orthogonal Arrays, Taguchi Method and Surface Finish.

INTRODUCTION

Surface finish is an important quality characteristic in many product design cases. The surface finish of a component is critical in situations where parts are to be assembled under tight tolerances or when the part is a moving component in a machine. Surface Roughness can also affect other part attributes like friction, reflectivity, heat transmission, lubricant retention, coating and fatigue. Surface finish in a milling operation depends upon several controllable and uncontrollable variables, such as spindle speed, feed rate, type of tool, tool material, work material machine tool vibration, tool chatter and environmental conditions. In Robust Engineering Using Taguchi Methods, these variables are separated into "control factors" and "noise (uncontrollable) factors". In this case, the Taguchi method has been used to study the effect of four control factors on the quality of surface finish. Results identify variables with the greatest effect on the critical quality characteristic.

SURFACE FINISH AND ITS MEASUREMENT

The precise measurement of surface texture irregularities of machined surfaces is difficult because the irregularities are complex in shape and character. Some of the characteristics that can be used to identify surface roughness are shape, length, luster, color, microstructure, waviness width, waviness height, roughness height and direction. Standards used for designating surface finish (ANSI B46.1, 1985) use roughness height, roughness width, waviness height, waviness width and lay as the critical characteristics, as shown in Figure-1. Various methods of measuring surface roughness are used in industry, such as average roughness (R_a), root mean square(rms) roughness (R_q), and maximum peak to valley roughness (R_y) etc. This paper uses the average roughness R_a. Average roughness was measured with a profilometer at four different locations. Average roughnes height is given by:

$$R_a = \frac{1}{L}\int_0^L |Y(x)|\,dx \qquad (1)$$

PROBLEM DESCRIPTION

Surface finish in a milling operation depends upon several variables, such as spindle speed, feed rate, type of tool, tool material and work material. A process planner traditionally chooses spindle speed and feed rate from the tables available in handbooks [1], [2]. These tables of machinability data are developed empirically and usually are given for maximum material removal rate conditions. However, maximum material removal rate is not always the critical quality characteristic while manufacturing a part. Surface roughness which is a measure of surface quality is often a major quality attribute of a manufactured part. In situations where the surface finish is a critical design specification, knowledge about relative effects of the above variables on the surface finish can lead to the selection of optimum cutting parameters. Knowledge about the relative influence of these cutting parameters on the surface finish can help machinists quickly and accurately achieve the specified surface finish. Lin [3] suggested a trial and error method to obtain optimum cutting parameters. However, this method is time consuming.

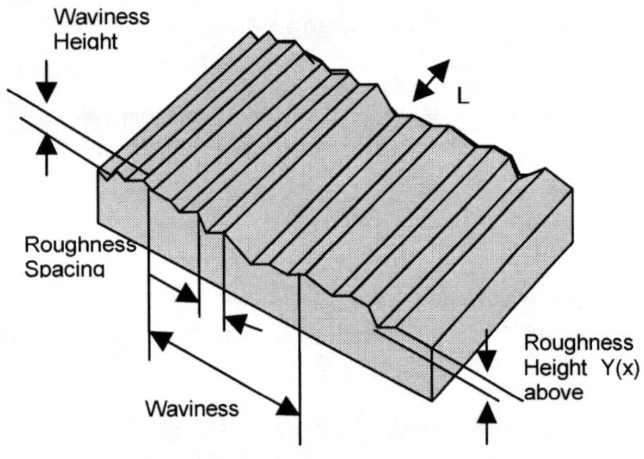

Figure-1, Surface Roughness Terminology

TAGUCHI METHODS

Sir R. A. Fisher first proposed the technique for design of experiments involving multiple factors. The method is popularly known as the factorial design of experiment (DOE). A full factorial design will include all possible combinations of a given set of parameters. When the number of parameters becomes large, however, the number of experiments required to perform full factorial design increases exponentially. Fridan, Kraft, Ruff and Derby [4] applied the DOE to investigate solder quality where they studied three variables at three levels.

$$Number\ of\ Experiments = n^k \qquad (2)$$

n = Number of Levels
k = Number of Parameters

To reduce the number of experiments to a practical level, Fisher and others designed a method based upon partial factorial design, using orthogonal arrays. This method provides for efficient exploration of design space with a minimum number of experiments. Dr. Genechi Taguchi improved on these methods to further improve the efficiency of answering the question of how to make a product or a process do what it is supposed to.

The Taguchi method uses a time-tested statistical technique to design experiments. Traditionally, engineers tend to change only one variable of an experiment at a time. The strength of Taguchi's technique is that multiple variables can be changed simultaneously without losing control of the experiment. It is usually not possible to include all of the variables in the design of experiments. In order to keep the experiment manageable, major variables and their nominal values are selected, based upon past experience and knowledge of the process. Then, in a prescribed manner, several variables are manipulated in an iterative, simultaneous fashion using orthogonal linear arrays. The effect of each variable is estimated in relation to critical quality characteristics, using a response table. The response is often affected by external noise conditions. To study the effect of the noise factors, Dr. Taguchi defined a term called signal-to-noise ratio [8]. In ideal combinations of control factor settings, where the response is highly resistant to the deteriorating affects of noise factors, the signal-to-noise ratio will be high. Therefore, in order to find the optimum level of cutting parameters, we need to maximize the signal-to-noise ratio. Analysis of variance is used to identify the relative importance of various factors on the response variable. Once the optimum levels for various parameters are identified, thereby providing the least variations closest to the target specification, a confirmation experiment is performed to verify results.

REVIEW OF LITERATURE

Taguchi methods have been applied to several areas in manufacturing and design. Lin and Kackar [5] looked at optimizing the wave soldering process. Mitchell [6] has researched the reliability of clearance defects on printed circuit boards. Stanley and Unal [7] have studied the optimization of dual mixture ratio for space propulsion systems. Phadke [8] provides an excellent review of the robust design methods. Krackar [9] discussed the Taguchi Method in detail. Wilkins [10] applied the Taguchi method to solve design problems with automotive wiper blades. Liu et al. [11] applied Taguchi methods to study the influence of various surface treatments on the performance of organic, light-emitting diodes. Published work on the application of Taguchi methods to the metal-cutting operation is limited. Lou, Chen and Lee [12] used multiple regression analysis to predict surface finish in the end-milling process. Yang and Chen [13] studied the surface finish in end milling aluminum by including four factors at three levels. Verma et al. [14] studied the surface finish on acrylic in the end milling process using a L4 orthogonal matrix. This study uses mild steel as the work material.

SELECTION OF AN ORTHOGONAL ARRAY

A number of standard orthogonal arrays are available. Each array can study a specific number of independent design parameters and levels. For example, if one wants to study the influence of four parameters, with each having three levels, then an L9 orthogonal array is the right choice. While a full factorial design will require 3^4 = 81 experiments, using L9 orthogonal array requires only 9 experiments. Table-1 shows some standard orthogonal arrays and the numbers of parameters and levels that can be studied with them.

No. of Levels	No. of Parameters	Orthogonal Array	Number of Experiments	No. of Full Factorial Exp.
2	3	L4	4	8
2	7	L8	8	128
3	4	L9	9	81
2	11	L12	12	2048
2	15	L16	16	32,768

Table-1, Some Standard Orthogonal Arrays

This paper discusses the influence of four parameters on surface roughness in end milling namely feed rate, cutting speed, depth of cut and tool type. L_9 orthogonal array can be used to study four parameters at three levels, as shown in Table-1. If we define the three levels as L (low), M (Med) and H (high), then Table-2 shows the various combinations of parameters and levels. A_L represents the value of surface roughness when parameter A has low value.

Parameters	Parameter Levels		
	L	M	H
A	A_L	A_M	A_H
B	B_L	B_M	B_H
C	C_L	C_M	C_H
D	D_L	D_M	D_H

Table-2, Parameters and Levels

SELECTION OF LEVELS & EXPERIMENTAL SETUP

Once parameters to be studied are identified, an appropriate level for these parameters must be selected. The parameter values must lie within the equipment capability, and at the same time the range should be large enough to provide a distinguishable effect on the quality characteristic namely, surface finish. The selected values for feed rate, spindle speed and diameter are given in Table-3. The experiment was performed on Kearney & Trecker Vertical Mill S-15. The constant parameters included tool diameter = 2 inch and width of cut = 1 inch. Figure –2 shows the experimental setup of the end milling process.

Parameter Assigned	Para-meters	Parameter Levels		
		L	M	H
Cutting Speed RPM	A	60.2 ft/min 115	87.96 ft/min 168	128.8 ft/min 246
Feed Rate	B	0.74 in./min	3.1 in/min	7.5 in/min
Depth of Cut	C	0.0625 in	0.125 in	0.25 in
Type of Tool	D	HSS	Carbide	Rough Cut

Table-3, Assigned Parameter Levels

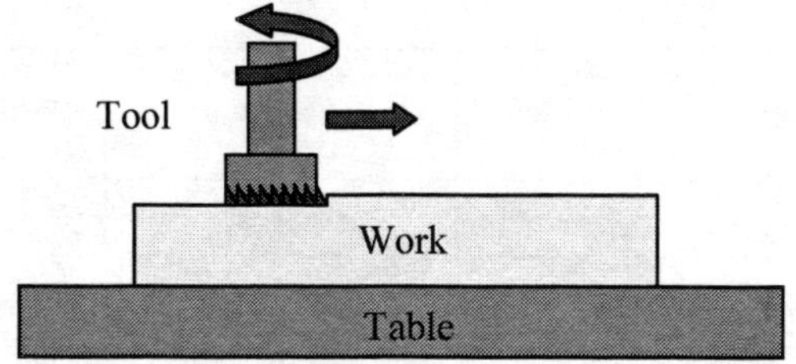

Figure-2, Experimental Setup

Table-4 shows the complete design of experiment using a L_9 orthogonal array.

Exp. No.	Actual Seq.	Cutting Speed Ft/min	Feed Rate In/min	Depth of Cut in	Tool Type	Avg. Roughness μ inch
1	4	60.21	0.74	0.0625	HSS	72.25
2	1	60.21	3.1	0.125	Carbide	55
3	7	60.21	7.5	0.25	Rough Cut	169.25
4	8	87.96	0.74	0.125	Rough Cut	100.75
5	6	87.96	3.1	0.25	HSS	155.75
6	2	87.96	7.5	0.0625	Carbide	41
7	3	128.8	0.74	0.25	Carbide	67
8	9	128.8	3.1	0.0625	Rough Cut	182
9	5	128.8	7.5	0.125	HSS	230

Table-4, Design of Experiment Setup

NOISE FACTORS

Noise factors affecting the end milling process may create variation in the quality of surface finish from point to point. Thus, location of measurement has been considered as a noise factor in this study. Surface measurements were taken at four different locations on the specimen. These results are shown in Table-5.

SIGNAL-TO-NOISE RATIO

The response variable considered in this study, Surface Finish, is of the smaller the better kind, that is, the lower the surface finish value the better. Therefore, signal-to-noise ratio is defined by the equation below [15]

$$\eta = -10\log_{10}\left[\frac{1}{n}\sum_{i=1}^{n} Y_i^2\right] \qquad (3)$$

where Y_i's are the individual measurement of roughness at each location.

CONDUCTING THE EXPERIMENT

The experiments were conducted on a vertical mill at Northrop Grumman Newport News. The end milling was done on mildsteel using three types of tools, high-speed steel end mills , carbide end mills and rough cut end mills. After the surface was machined, roughness was measured using a microscope. Roughness height measurements were taken using a profilometer. Four readings for each specimen were obtained, and the average was calculated, as shown in Table-5. The values of roughness ranged from 35 to 249 micro inches.

Exp #	Loc.–1 μ inch	Loc.–2 μ inch	Loc.–3 μ inch	Loc.–4 μ inch	Avg. Roughness μ inch
1	70	85	69	65	72.25
2	53	45	58	64	55
3	171	165	179	162	169.25
4	87	98	119	99	100.75
5	157	153	151	162	155.75
6	40	42	47	35	41
7	64	65	69	70	67
8	183	176	194	175	182
9	203	244	249	224	230

Table-5, DOE Data

ANALYSIS OF MEANS

Table-6 shows the response table or the main effect table of the data shown in Table-5. Main effect of cutting speed at level 1 is equal to average of surface roughness of all the experiments with cutting speed at level 1. Thus, the main effect of the cutting speed at level 1 is MA_1 = (72.25+55+169.25) /3 = 98.8. Delta represents the maximum change in the value of surface roughness between the three levels. The large value of delta represents a larger influence on the roughness. Delta corresponding to type of tool is the largest followed by feed rate, cutting speed and depth of cut and are ranked accordingly from 1 to 4.

Level	A Cutting Speed	B Feed Rate	C Depth of Cut	D Type of Tool
1	98.8	80	98.4	152.66
2	99.16	130.91	128.58	54.33
3	159.66	146.75	130.66	150.66
Delta Δ	60.86	66.75	32.26	98.33
Rank	3	2	4	1

Table-6, Mean Response Table

RESPONSE GRAPHS FOR MEAN

The values obtained from the response table are plotted to visualize the effect of the three parameters. Figure-3 shows the influence of cutting speed (A), feed rate (B), depth of cut (C) and type of tool (D) on average surface roughness. As feed rate increases, surface roughness increases.

As tool depth of cut increases, surface roughness increases. In the graph above D1 corresponds to HSS tool, D2 corresponds to carbide tool and D3 corresponds to rough cut mill. Surface roughness was found to be lowest for carbide tool.

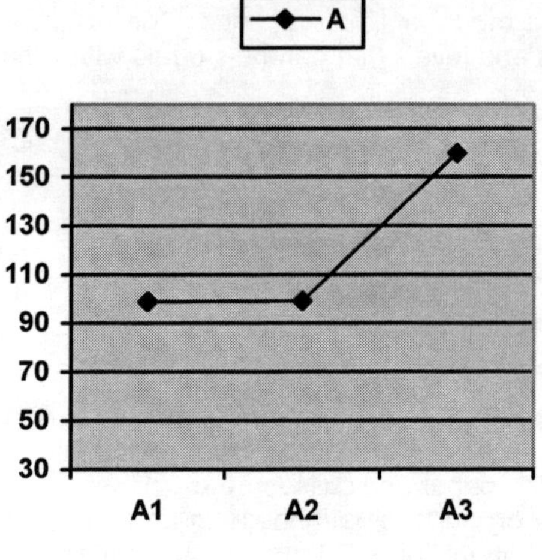

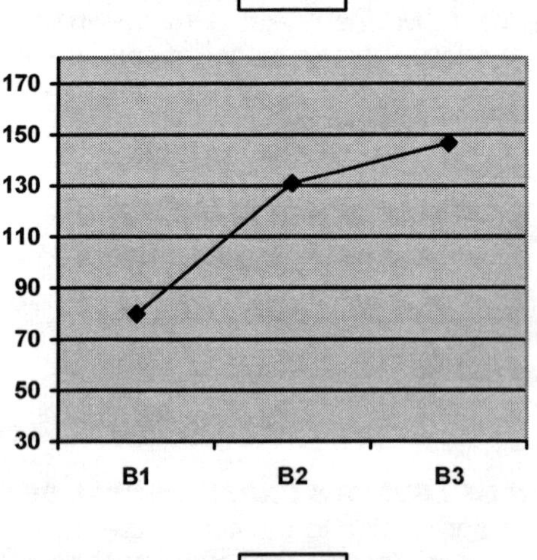

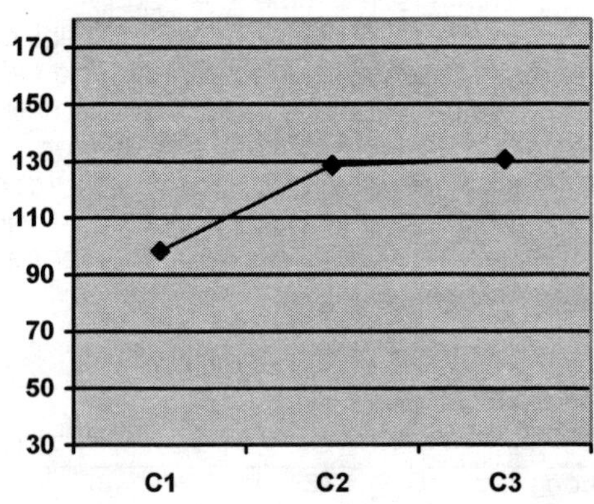

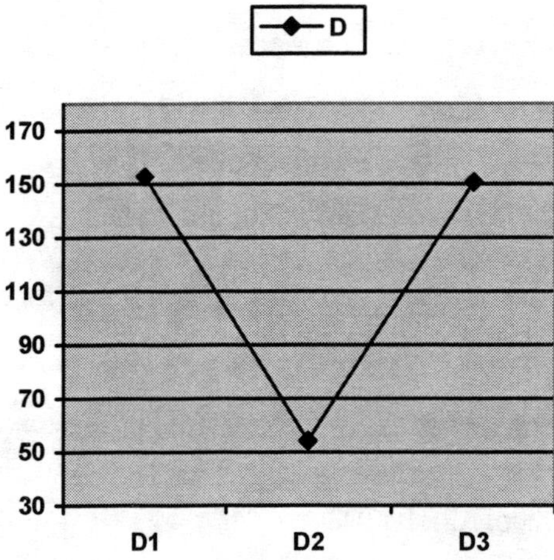

Figure-3, Mean Response Graphs

ANALYSIS OF SIGNAL-TO-NOISE RATIO

The Signal-to-Noise ratio analysis evaluates average surface roughness as well as variation around the average in a single numerical value. A response table for signal-to-noise ratio is generated, as shown below in Table-7. These values are plotted in Figure-5, and the optimum levels are circled. The parameter levels that give us maximum signal-to-noise ratio are A1 or 2, B2 and C2, which match the levels for the minimum surface roughness, as obtained in the analysis of means. The signal-to-noise ratio analysis further confirms our earlier finding that, among the three factors studied, type of tool has the maximum influence on the surface roughness (delta = 8.73) with spindle speed having the second strongest affect (delta = 4.24). If Analysis of Means gave very different results from Signal-to-Noise Ratio analysis, we would use the factors with the greatest affect on Signal-to-Noise Ratio to reduce variation in surface roughness first. We would then use one factor with the greatest affect on Analysis of Means results to adjust the surface roughness to the desired level.

Level	A Cutting Speed	B Feed Rate	C Depth of Cut	D Type of Tool
1	-38.89	-37.96	-38.25	-42.78
2	-38.76	-41.31	-40.75	-34.57
3	-43	-41.38	-41.65	-43.30
Delta Δ	4.24	3.42	3.41	8.73
Rank	2	3	4	1

Table-7, Signal to Noise Ratio Response Table

Figure- 4, shows the response graphs for signal to noise ratio.

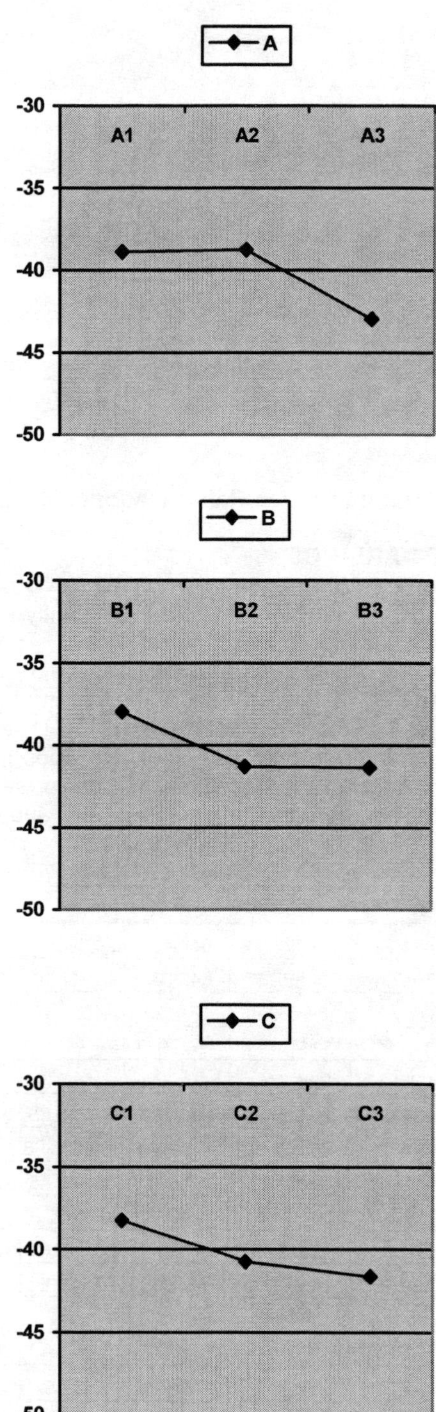

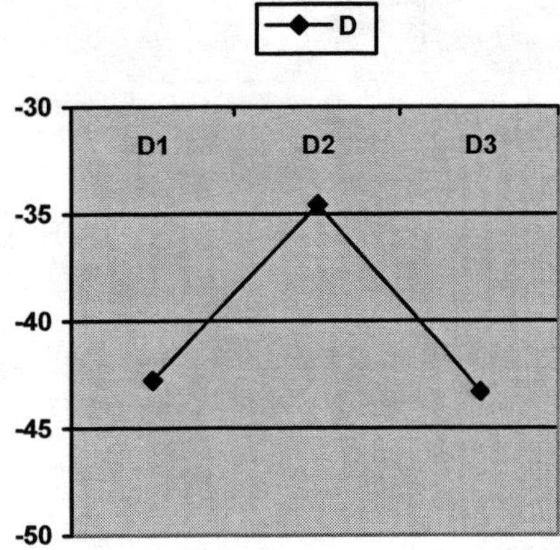

Figure-4, Signal to Noise Ratio Response Graphs

ANALYSIS OF VARIANCE

The purpose of analysis of variance (ANOVA) is to determine which cutting parameters significantly affect the quality characteristic, in this case roughness. Table 8 shows the results of ANOVA analysis. The value of F factor is a predictor of how influential the parameter is. Since the F value for feed rate (C=17.13) is small compared to others (A= 64.5, B= 64.0, D= 166.2), we eliminate the effect of c in calculating the predicted optimum value. The interaction effects AxB, AxC and BxC are large.

Source	DF	S	V	F	S'	rho
A	2	29,444.22	14,722.11	64.5497	28,988.07	19.98
B	2	29,195.06	14,597.53	64.0035	28,738.91	19.8
C	2	7,817.72	3,908.86	17.1386	7,361.57	5.07
D	2	75,814.22	37,907.11	166.2053	75,358.07	51.93
P	3	684.2222	228.0741			
A x B	4	83,631.94	20,907.99	91.6719	82,719.65	57
A x C	4	105,009.28	26,252.32	115.1044	104,096.98	71.73
A x D	4	37,012.78	9,253.19	40.571	36,100.48	24.88
B x C	4	105,258.44	26,314.61	115.3775	104,346.15	71.9
B x D	4	37,261.94	9,315.49	40.8441	36,349.65	25.05
C x D	4	58,639.28	14,659.82	64.2766	57,726.98	39.78
e1						
e2						
(e)	3	684.2222	228.0741		7,982.59	5.5
Total	35	145,120.22	4,146.29			

Table-8, Analysis of Variance Table

PREDICTION & CONFIRMATION

Using the data obtained above for the strongest control factors, one could predict the optimum surface roughness using the cutting parameters identified.

$$\hat{\mu} = \overline{\mu} + (\overline{\mu} - \mu_D) + (\overline{\mu} - \mu_A) + (\overline{\mu} - \mu_B) \qquad (4)$$

$\overline{\mu}$ = Mean surface roughness

μ_A = Main effect due to A

μ_B = Main effect due to B

μ_D = Main effect due to D

The results for mean are shown in table – 9 and for signal to noise ratio is shown in table-10.

Mean Surface Roughness		
	Predicted	Actual
Initial A2B1C2D3	111.44	100.75
Optimum A2B1C1D2	15.11	12.00
Gain	96.33	88.75
Variation Reduction	86%	88%

Table-9, Prediction and confirmation results for mean

S/N Ratio		
	Predicted (dB)	Actual (dB)
Initial A2B1C2D3	-41.84	-40.12
Optimum A2B1C1D2	-33.11	-21.60
Gain	8.73	18.52
Variation Reduction	64%	88%

Table-10, Prediction and confirmation results for S/N ratio

This study confirms well for average surface roughness results, but not so well for S/N ratio, which evaluates variation in the surface roughness. Our actual S/N ratio results achieved were dramatically better (24% better variation reduction) than our prediction. While this is good news, we cannot be sure why the results are so much better (i.e. interaction effects, experimental error, measurement error, an unexpectedly strong noise factor present, etc) without further study. As a result, we would implement these optimum settings into production with some degree of caution and carefully watch the production end milling results for a while to make sure we did not get any surprises.

CONCLUSIONS

The effect of four cutting parameters has been studied on the quality of surface finish in the end-milling of mild steel. Results indicate that the type of tool has the maximum influence on average surface finish, as indicated in the response tables for mean (Table-6) followed by feed rate, cutting speed, and depth of cut.

Signal-to-noise ratio analysis (Table-7) indicates that type of tool, cutting speed, feed rate and depth of cut are influential on surface variation in the respective order. Predicted optimum values for surface roughness correspond well with the average surface roughness results of the confirmation experiment and gave better than expected results for surface variation (signal-to-noise ratio). Actual reduction in surface finish as a result of the experimental analysis was 88%.

ACKNOWLEDGMENTS

This study was done at Northrop Grumman Newport News. The authors are indebted to Dr. James Hughes for facilitating the study and to the Apprentice School for use of their machine shop.

REFERENCES

1. Machinery's Handbook, 23rd revised edition, Industrial Press, NY, 1988.
2. Machinability Testing and Utilization of Machining Data, American Society of Metals, 1979.
3. Lin, S. C., "Computer Numerical Control- from Programming to Networking." Delmar, New York, 1994.
4. Fridan, I., Kraft, R. P., Ruff, L. E. & Derby, S. J. "Designed Experiments to Investigate the Solder Joint Quality Output of a Prototype Automated Surface Mount Replacement System." IEEE Transactions, V21, No. 3, p 172-181.
5. Lin, K. M. and Kackar, R. N., "Wave Soldering Process Optimization by Orthogonal Array Design Method." Electronic Packing and Production (Feb. 1985) pp. 108-115.
6. Mitchell, J. P. "Reliability of Isolated Clearance Defects on Printed Circuit Boards." Proceedings of Printed Circuit World Convention IV (June 1987), Tokyo, Japan, pp. 50.1-50.16.
7. Stanley, D. O. and Unal, R. "Application of Taguchi Methods to Dual Mixture Ratio Propulsion System Optimization for SSTO Vehicles." 30th Aerospace Sciences Meeting & Exhibit, January 6-9, 1992, Reno, Nevada.
8. Phadke, S. M., "Quality Engineering Using Robust Design," Prentice Hall, Englewood Clifffs, New Jersey, 1989.
9. Kackar, R. N., "Taguchi's Quality Philosophy: Analysis and Commentary." Quality Progress (Dec. 1986) pp. 21-29.
10. Wilkins, James O., "Putting Taguchi Methods to Work to Solve Design Flaws." Quality Progress, May 2000.
11. Liu, Jia-Ming, Lu, Po-Yen, Weng, Wen-Kou, "Studies on Modification of ITO Surfaces in OLED Devices by Taguchi Methods."
12. Lou, Mike S., Chen, Joseph C., Li, Caleb M., "Surface Roughness Prediction Technique for CNC End-Milling." Journal of Industrial Technology, Vol. 15, No. 1, November, 1998.
13. Yang, J. and Chen, J., "A Systematic Approach for Identifying Optimum Surface Roughness Performance in End Milling Operations." Journal of Industrial Technology, Vol.17, No. 2, 2001.
14. Verma, A. K. and Holcomb, S. L., "Applying Taguchi Method to Parametric Study of Surface Finish in End Milling." International Journal of Agile Manufacturing, Volume 7, Issue 2, 2004.
15. Taguchi, Genichi, Chowdhury, S. & Taguchi, Shin, "Robust Engineering", McGraw Hill, 2000.

Proceedings of IMECE'03
2003 ASME International Mechanical Engineering Congress
Washington, D.C., November 15–21, 2003

IMECE2003-42306

USING TUNED MASS DAMPERS TO SILENCE A COORDINATE MEASUREMENT MACHINE

David H. Myszka
University of Dayton

ABSTRACT

Nearly every manufacturing operation relies on servo-controlled automation and inspection machines. The design of these machines requires close attention to the relationships of the mechanical system with the electronic control system. Often, the attributes of one component can augment the weaknesses of another, and vice versa.

For example, linear motions are traditionally produced with a rotary motor and ball screw. Recently, linear motors have gained popularity because of their numerous advantages. However, the servo-controls can, more easily, excite resonant frequencies in the machine structure. Therefore, controlling the machine at various speeds can produce excessive structural vibrations as different modes are excited. Thus, improvement in one component causes difficulties with another.

An efficient solution is a tuned mass damper (TMD). This is a simple, modular device that consists of a weight, mounted on a spring along with a viscous energy absorber. The tuned mass damper is placed where the undesired vibratory motion is greatest. The size of the mass, spring and damper are adjusted, or tuned, so they oscillate out-of-phase with the structure and reduce the amplitude of vibration.

This paper will review the theory of tuned dynamic dampers, and illustrate an application of integrating them into the design of a coordinate measuring machine.

NOMENCLATURE

K Structure stiffness
M Structure modal mass
k TMD stiffness
m TMD mass
f Natural frequency

INTRODUCTION

Tuned mass dampers (TMDs) can be a practical and effective manner to reduce resonant vibration in structures [1, 5, 9, 10]. Shown schematically in Figure 1, a TMD is a modular device composed of a spring, mass, and damper.

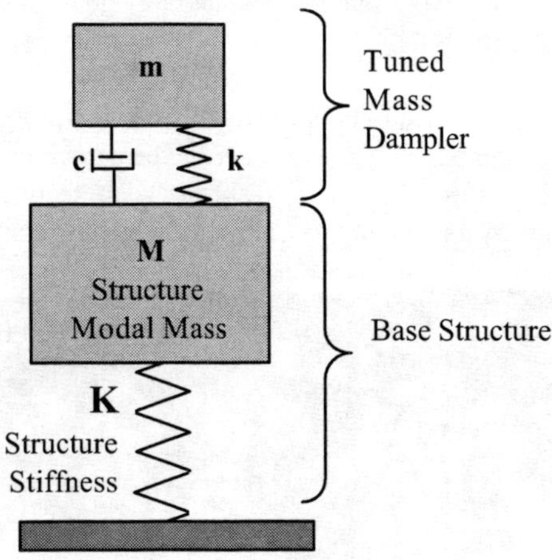

Figure 1

A TMD is a single degree-of-freedom mechanical system, attached to a base structure. It adds a mode of vibration to that base structure. Mounted to a rigid base, the TMD would have a natural frequency of:

$$f = \frac{1}{2\pi}\sqrt{\frac{k}{m}} \qquad (1)$$

In the design of the actual TMD device, the stiffness k and mass m are chosen to put the TMD natural frequency just below the frequency of the "target" mode of the base structure. This causes a strong dynamic interaction between the TMD and the target mode. The target mode is replaced by two modes, one slightly above and one slightly below the original frequency. Most important, both of these "split modes" will be damped by the dashpot of the TMD. In essence, the TMD attracts the vibrational energy of the target mode and dissipates it into heat through the action of its dashpot.

Much research has been done on using tuned mass dampers in civil engineering applications such as bridges, smoke stacks, and high-rise buildings [14, 17]. Some work has been done in machine tool applications [12, 13, 16]. A unique application arose with a measurement machine with vibration problems. This paper discusses applying this technology on ultra-precision equipment.

CNC-CMMS

Computer numerical controlled, coordinate measuring machines (CNC-CMM) have become popular in the inspection of high volume parts. The specific inspection features of a part can be programmed. Then, the servo-controlled axes of the CNC-CMM position a probe to move toward and along the desired features. Specialized software is available to readily measure complex features, such as gear teeth, splines, sprockets, etc.

The geometric data is captured from the probe and stored. This data can be analyzed to describe the actual measurement of a single feature, the relationship between features, or trends over time. A representative CNC-CMM is illustrated in figure 2.

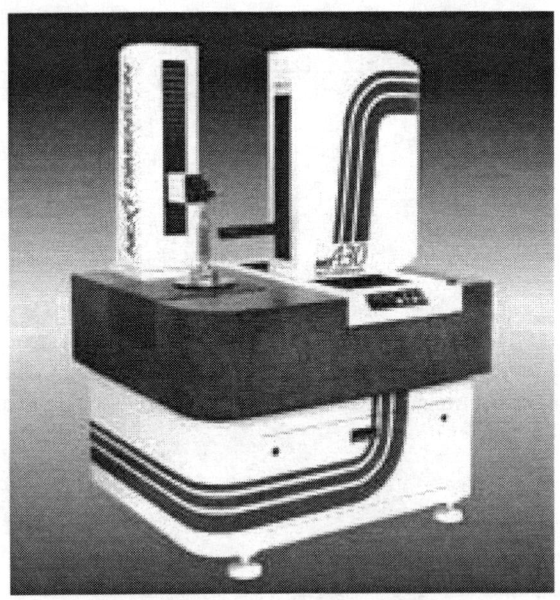

Figure 2

Typically, these measurement machines are thermally and volumetrically mapped and error compensated. Sub-micron accuracy is achievable. With such a precise and delicate machine, the use of mass dampers was considered impracticable.

CNC-CMM DESIGN

The electromechanical design of most CNC-CMMs is similar. The three mobile axes are arranged in a moving ram, horizontal arm configuration. This arraignment is shown in figure 3.

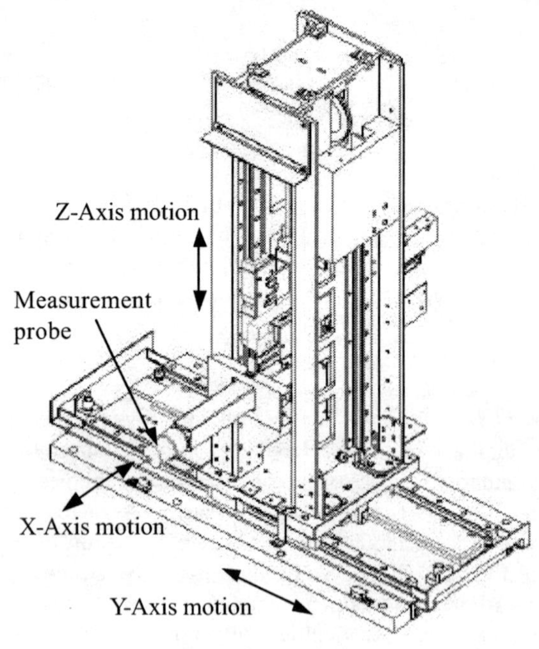

Figure 3

The structure is lightweight, typically cast, or bolted plate, aluminum. The vertical, Z, axis is balanced with a counterweight. Each axis is supported with commercial linear bearing systems. The axes are driven with servomotors. Linear encoders are attached to the base of the axes, to provide positional feedback.

Lightweight materials and a counterbalance are used to minimize the power required from the servomotors. Minimizing requirements from the motors, reduces heat and thermal growth in the structure. Thermal effects are extremely detrimental to a measurement machine. Additionally, thermal hysteresis from bolted connections and non-linear materials are severely affect measurement repeatability. For this reason, use of damping material is not viable.

SERVOMOTOR DRIVES

Traditionally, the linear axes of the CNC-CMMs were driven with rotary servomotors coupled with ball screws. The trend in the past few years has been to use linear motors, eliminating the need for a ball screw to act as a rotary to linear converter.

Without the conventional drive train, the linear drive is also without the drive train's inherent backlash and compliance, issues that increase position uncertainty. Thus, the linear motors appear ideal for high speed, high precision movements. Another major advantage is that without surfaces in physical contact, mechanical wear is eliminated, achieving quiet operation. Last, but not least, the prices of linear motors are still falling.

SERVO INDUCED RESONANCE

Every machine consists of various assemblies, which tend to oscillate because they basically are mass-spring systems. They have a resonant frequency that results from rigidity or stiffness of the mechanical machine members and their mass. The resonant frequency, in turn, limits the maximum acceleration rate that the machine can achieve.

Because of its increased responsiveness, the servo-system of a CNC-CMM with linear motors can more easily excite resonant frequencies in the machine structure. Therefore, controlling the machine at various speeds can produce excessive structural vibrations. This is particularly true in CNC-CMMs since damping materials are not used because they tend to cause thermal hysteresis.

For the CNC-CMM considered, a proportional-integral-differential (PID) control scheme is used [16. However, this machine has a lightweight structure and minimal damping. Consequently, instability resulted when the servo gains were raised to levels necessary to achieve desired performance. As the gain was increased, performance was steadily improving until the machine becomes unstable, blaring with an unpleasant sound, similar to a fog horn. It was nearly impossible to tune the system at the required speed without exciting a resonant frequency.

MODAL ANALYSIS

Modal analysis determines the dynamics characteristics of a vibrating system from experimental transfer functions. The dynamic characteristics include the mode shapes and the modal parameters. A mode shape is the deformation shape at a single natural frequency. The modal parameters are the natural frequencies, damping ratios, and modal masses associated with each of the mode shapes.

A transfer function is a measure of the response of a system to a given input. In vibration analysis, the input is usually a force and the response measured is usually a motion, either displacement, velocity, or acceleration. In order to obtain a transfer function, a known force is applied at one point and the acceleration, for example, is measured at another point. The transfer function can then be obtained by taking the ratio of the two measurements.

For a complex system, the transfer functions are usually obtained with specialized software packages. The input and response measurements are digitized and then processed using fast Fourier transform techniques. The transfer function is calculated and stored as a finite number of data points. Once the transfer functions are obtained, the natural frequencies and mode shapes are obtained. From this information, "pictures" of how the system is vibrating at its different natural frequencies can be determined and animated for better visualization.

For the CNC-CMM, an accelerometer was used at 18 points, defining the vertical structure and the probe arm. Modal analysis was performed using STARModal software. A natural frequency was discovered at 230 hz, with extremely low damping. This frequency was audibly consistent with the growling sound of the machine. The corresponding mode shape is illustrated in figure 4.

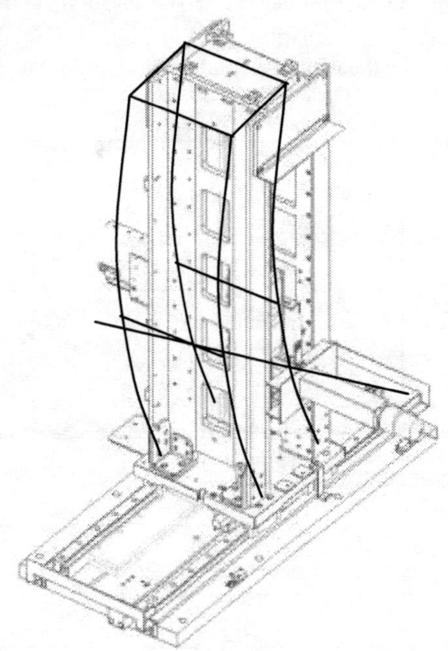

Figure 4

DESIGN OF THE TMD

Although it has never been documented, it was decided to attempt using a TMD on this ultra-precision measurement machine. This involves selecting the TMD mass, m, stiffness, k, and the damping, c.

Figure 5 illustrates the TMD design. The TMD is attached to the machine structure at the location with the greatest vibrational displacement for the mode to be dampened.

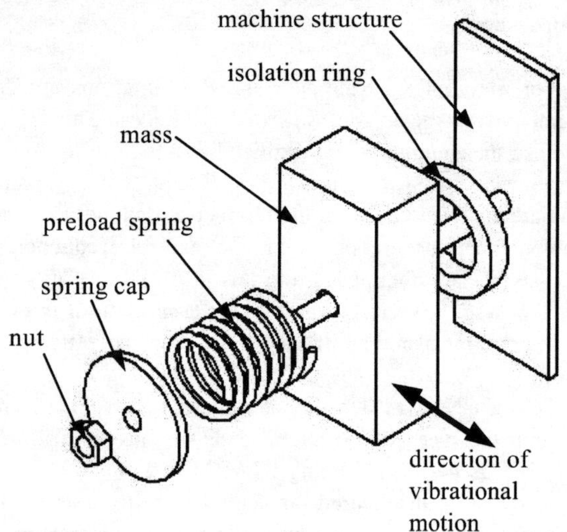

Figure 5

A general TMD rule of thumb is to select m at 5-10% of M. With a vertical structure weight of 180 lbs, a 10 lb TMD mass was selected. Using equation 1, a stiffness of 540,000 lb/in is required to place the natural frequency of the TMD to 230 hz.

Several different isolation rings were considered. AeroFlex, E-A-R, Lord, and Berg sell isolation rings as standard products. Since the rings have a non-linear stiffness, a ring was selected that had a transverse stiffness close to that required. A preload spring and nut are used to load the ring, and "fine tune" the TMD stiffness.

The process of adjusting the preload, i.e., fine tuning, was somewhat trial and error. A frequency plot of the signal from an accelerometer was observed after impacting the structure. The nut was tightened until the target frequency was removed.

RESULTS

Figure 6 is a frequency plot from an accelerometer after an impact strike on the machine structure prior to attaching the TMD. Notice the large amplitude at the 230 hz frequency. Figure 7 is another frequency plot from an accelerometer after an impact strike on the machine structure after "fine tuning" TMD. Obviously, the resonance at the 230 hz frequency has been virtually eliminated.

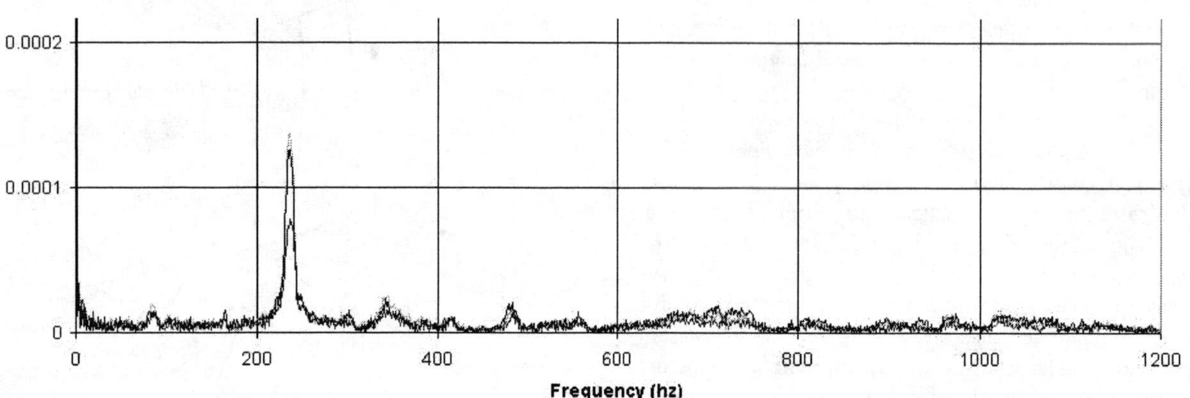

Figure 6

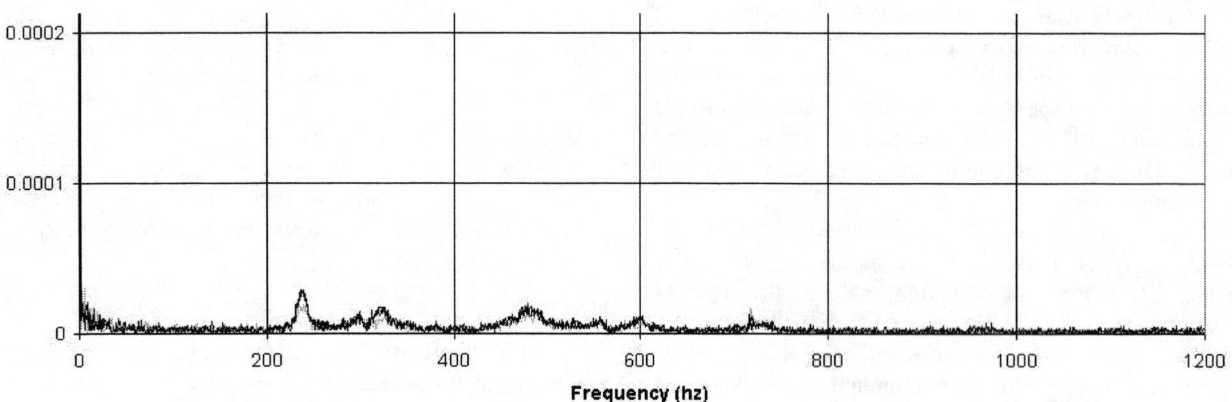

Figure 7

The ultimate result of the TMD is that the P-I-D gains could be adjusted to allow the machine to achieve the desired speeds. The machine was extremely stable, and the gains needed to be grossly altered to set the machine into a resonant vibration. It should be noted that this vibration was a different frequency and mode. If this resonance were a problem, an additional TMD could be designed for the other mode.

CONCLUSIONS

As a solution to a vibration problem, TMDs have a number of very attractive features:

- ✓ They are inherently compact, modular devices that can have a simple interface to the base structure.
- ✓ They can be readily added to a base structure that is already designed or even built.
- ✓ A well-designed TMD can add high damping with minimal weight.
- ✓ The TMD does not impact the static strength or stiffness of the base structure.
- ✓ For designing the TMD, it is often possible to characterize the base structure by inexpensive test or analysis. In effect, the base structure is modeled simply in terms of a single mode: the target mode.
- ✓ A TMD is an efficient solution for ultra-precision, machine vibration problems.

REFERENCES

1. DIAS V. Tuned Dampers Improve Stepper Performance, *E-A-R Division, Cabot Corporation.*
2. FLINT E., EVERT M., ANDERSON E., FLANNERY P., 2000. Active/Passive Counter-Force Vibration Control and Isolation Systems, *Proc. of IEEE Aerospace Conference.*
3. GARDNER R., The Reduction of High Frequency Vibration and Noise by Using E-A-R, *E-A-R Division, Cabot Corporation.*
4. HARRIS C., 1991. *Handbook of Acoustical Measurements and Noise Control*, McGraw-Hill.
5. INAUDI J., KELLY J., 1995. Mass Damper Using Friction-Dissipating Devices, *Journal of Engineering Mechanics,* 121/1.
6. INMAN D., 1994. *Engineering Vibration*, Prentice-Hall.
7. MATERSON P., The Basics of Vibration Isolation using Elastomeric Materials, *E-A-R Division, Cabot Corporation.*
8. MEIROVITCH L., 1975. *Elements of Vibration Analysis*, McGraw-Hill.
9. PAZ M., 1997. *Structural Dynamics: Theory and Computation"* Chapman & Hall.
10. RANDALL S. E, 1981. Optimum Vibration Absorbers for Linear Damped System. *Journal of Mechanical Design*, Transactions of the ASME, 908/ 103.
11. RENNINGER J., Understanding Damping Techniques for Noise and Vibration Control, , *E-A-R Division, Cabot Corporation.*
12. RIVEN E., KANG H., 1989. Improvement of Machining Conditions for Slender Parts by Tuned Dynamic Stiffness of Tool, *Int. J. Mach. Tools Manuf.*, 29/3.
13. SLOCUM, A., 1992. *Precision Machine Design*, Society of Manufacturing Engineers.
14. WARBURTON G., 1982. Optimum Absorber Parameters for Various Combinations of Response and Excitation Parameters, *Earthquake Engineering and Structural Dynamics,* 10.
15. WELCH R., 2002. Mechanical Resonance in a Closed-Loop Servo System, *Tutorial from Welch Enterprises.*
16. WERTZ H, 2000. Self-tuning Speed Control for Servo Drives with Imperfect Mechanical Loads, *Proc of IEEE IAS.*
17. YOUSSEF N., 1994. Supertall Buildings with Tuned Mass Damper, *Structural Design of Tall Buildings*, 3/1.

Proceedings of IMECE'03
2003 ASME International Mechanical Engineering Congress
Washington, D.C., November 15–21, 2003

IMECE2003-42312

COLLABORATIVE DEVELOPMENT OF A STRUCTURAL DAMPER UTILIZING ULTRA-HIGH-DAMPED ELASTOMER

Shannon K. Sweeney
Penn State Erie, The Behrend College
School of Engineering and Engineering Technology
Erie, Pennsylvania 16563-1701

Robert J. Michael
Penn State Erie, The Behrend College
School of Engineering and Engineering Technology
Erie, Pennsylvania 16563-1701

ABSTRACT

This paper defines a current applied research project in mechanical engineering technology, describes the roles of the principles, provides a status report of year one work, and provides the major goals for year two. The purpose of the project is to develop a marketable viscoelastic structural damper utilizing ultra-high-damped elastomer to protect buildings from seismic and wind events. The project is a collaboration of Penn State Erie, Lehigh University, and Lord Corporation that is funded by a grant from the Pennsylvania Department of Community and Economic Development through the Pennsylvania Infrastructure Technology Alliance.

INTRODUCTION

Extensive structural research and analytical modeling by Lehigh University [1], [4], [7] indicates that elastomeric dampers, having viscoelastic properties, can protect buildings from seismic and wind events as well as or better than the widely used viscous dampers and at a lower cost. Through their research, [2], [3], [5] Lehigh University has proposed using the technology of viscoelastic properties to control structural motions. This paper will outline the applied aspects of the research that are currently underway to develop and advance this technology in the seismic damper market.

The purpose of the project is to develop materials, to verify analytical models, to investigate the commercial feasibility of elastomeric structural dampers, and to support the product quality planning process. The goal of the project is to drive the concept of elastomeric structural dampers from the idea stage through product development, market launch and beyond.

The applied aspects of this research project, including coordination responsibility, have been assigned to the Mechanical Engineering Technology program at Penn State Erie. As a result, Penn State Erie has conducted the background research, will design the damper material and configuration with Lehigh University, will coordinate material development and product testing with Lord Corporation, and will investigate the commercial potential for elastomeric dampers in the applicable market(s). Comments on the collaboration efforts are provided along with updates on material development, damper design, damper prototype, and product test preparations.

The Principles And Their Roles

The mission of the Pennsylvania Department of Community and Economic Development is to foster opportunities for businesses and communities to succeed and thrive in a global economy, thereby enabling Pennsylvanians to achieve a superior quality of life. The mission is accomplished by budgeting for enhanced business financing, expanded community development, technology advancement initiatives, and retention of young talent. This applied research project is an initiative to advance existing technology.

The Pennsylvania Infrastructure Technology Alliance (PITA) is a collaboration among the Commonwealth of Pennsylvania (through the Pennsylvania Department of Community and Economic Development), the Center for Advanced Technology for Large Structural Systems (ATLSS) at Lehigh University, and the Institute for Complex Engineered Systems (ICES) at Carnegie Mellon University (CMU). The mission of the PITA Research Program is to assist the Commonwealth of Pennsylvania and its companies in increasing operating efficiency and enhancing economic development.

Lehigh University plays an important role in the cultural and economic development of the greater Lehigh Valley region in which it resides. Lehigh students, faculty and staff interact with local residents on a regular basis and share their talents off-campus, often through research projects. The Center for Advanced Technology for Large Structural Systems (ATLSS) at Lehigh University is a center devoted exclusively to the construction and large-structure industries. The role of ATLSS is to conduct research and educate students on technology issues affecting these industries in design, fabrication, construction, inspection, and protection.

Lehigh University is a recognized research leader in the field of large-structure dynamics and has done extensive analytical modeling of energy dissipation devices used to protect large structures from seismic and wind events. See References [1] and [2] for leading examples. As a result, Lehigh University has provided the material property goals and the configuration requirements for the dampers. Lehigh University will later test the dampers in an applicable structure.

Lord Corporation, a privately owned company, originated in Erie, Pennsylvania and was founded in 1924 as Lord Manufacturing Company. Today, with annual sales over $400 million, the Corporation designs, manufactures and markets chemical and mechanical products worldwide. For more than 75 years Lord Corporation has been launching innovative products in the global marketplace. Those innovations include performance-formulated products for coating, bonding, and product assembly, and performance-designed components and systems for controlling vibration, shock, motion and noise.

Lord Corporation is a recognized leader in the field of elastomeric material development capabilities. As a result, Lord Corporation will develop an elastomeric material to the necessary property goals, provide material test data, select manufacturing processes, and build prototype damper(s).

As a Land Grant and Sea Grant institution, Penn State Erie, The Behrend College is uniquely positioned to advance Northwestern Pennsylvania's economic and social welfare. Students engage in the college's mission of teaching, research, and service by regularly participating in applied research and service projects.

The School of Engineering and Engineering Technology at Penn State Erie, specifically faculty within the mechanical engineering technology program with expertise in elastomer product and manufacturing design, has responsibility for overall project coordination. Penn State Erie has conducted the background research, will design the damper with Lehigh University, and will coordinate material and product testing with Lord Corporation. The School of Engineering and Engineering Technology will also work with the School of Business at Penn State Erie to assess market potential and develop a business plan for structural dampers.

Background: Philosophy and Technology

Both viscous dampers and elastomeric dampers utilize the technical philosophy of energy dissipation to control motion of a structure. A distinctly different technical philosophy for controlling motion is the utilization of a base isolation system. A major advantage of the energy dissipation philosophy is that existing structures can be more easily retrofitted with energy dissipation devices than with base isolation systems. This research project will not address other advantages or disadvantages of the two distinctly different technical philosophies for controlling motion.

Within the technical philosophy of energy dissipation, there are four main product types; viscous dampers, elastomeric dampers, friction dampers, and metal-yielding devices. The effectiveness of each is, at this time, an open debate. Currently, viscous dampers are the most widely used. Rigorous nonlinear time history analyses conducted by Lehigh University [6] show that elastomeric dampers are as effective at controlling structural motion as viscous dampers. Advantages of elastomeric dampers over viscous dampers are that they are potentially more cost-effective and that multi-directional damping is possible. A disadvantage of current elastomeric dampers is that the energy dissipation properties of the materials are sensitive to frequency, temperature, strain amplitude, and time, and may not perform as expected during seismic and wind events. Friction dampers and metal-yielding devices have the disadvantage of being all on or all off and therefore may not dissipate any energy during small structural motions.

Lord Corporation has been involved with large structural systems over many years. The area of building isolation has been in existence since the early 1940's for Lord Corporation and consists of using laterally soft springs to isolate buildings from earthquake (ground) vibration. Lord Corporation designed a series of elastomeric isolation bearings with load ranges of 100 kip up to 500 kip. These isolation bearings utilized several steel shims to produce high stiffness in the vertical direction (building natural frequency > 10 Hz), while maintaining soft lateral rates (typically 0.25 Hz or less) to accommodate displacements of up to +/- 3 inches.

Lord Corporation has also been involved with structural dampers or energy dissipation devices throughout the years. Recent effort includes applying magnetorheological (MR) fluid technology within active or semi-active dampers. This technology has been successfully used in large civil engineering structures such as the newly constructed cable-stayed bridge in Hunan Province, China. Throughout the years, Lord has also been engaged in various structural damper projects utilizing lower cost elastomeric and surface effect (SE) dampers. Many structural damper projects have been successful from a development and technical standpoint.

Based on limited marketing studies, active MR dampers in building structures may not offer enough of a performance benefit to justify the added cost and/or reduced reliability. According to tests performed at University of Buffalo, active control offers only a 15 – 20% performance improvement over a properly designed passively damped structure. Consequently, target markets for MR dampers tend to be performance driven military applications and special projects such as the Hunan Province Bridge.

Data for the five-year usage of traditional viscous dampers has been accumulated from two key manufacturers and summarized. The data shows that approximately 90% of all seismic dampers sold during this time period were at 337 kip (1,500 kN) damping force or less. This was important since the team felt that elastomeric dampers in this range were technically feasible. Based on Lehigh's analytical models, the good probability of developing an elastomer that is insensitive to dynamic and environmental conditions, and the apparent technical feasibility of elastomeric dampers for structures, the planned research of this project is merited. Further, the scope of the project makes it a good fit as an applied research program in Mechanical Engineering Technology.

COLLABORATION

A condition for funding of research by the Pennsylvania Infrastructure Technology Alliance is that Pennsylvania higher education and Pennsylvania industry be linked by the project. Support from industry may be in-kind. As a result, collaboration between Penn State Erie, Lehigh University, and Lord Corporation on this project has been and will continue to be extensive. The project will leverage the elastomeric product design and manufacturing expertise of Penn State Erie faculty, the modeling and structural expertise and facilities of Lehigh University, and the material development and testing expertise and facilities of Lord Corporation.

	Collaboration		Year 1				Year 2			
Action	Primary	Support	Qtr. 1	Qtr. 2	Qtr. 3	Qtr. 4	Qtr. 1	Qtr. 2	Qtr. 3	Qtr. 4
Identify Material Property Goals	Lehigh U.	PSU-Erie								
Preliminary Material Development	Lord Corp.	PSU-Erie								
Research Relevant Technologies	PSU-Erie	Lord Corp.								
Research Past Marketing Issues	PSU-Erie	Lord Corp.								
Preliminary Design Concept	Lehigh U.	PSU-Erie								
Finalize Material Development	Lord Corp.	PSU-Erie								
Product Design and Development	PSU-Erie	Lord Corp.								
Prototype Damper Build	Lord Corp.	PSU-Erie								
Process Design and Development	Lord Corp.	PSU-Erie								
Product and Process Validation	Lehigh U.	PSU-Erie								
Commercial Feasibility Study	PSU-Erie	Lord Corp.								

Table 1

Penn State-Erie has provided the part-time resources of two full-time faculty to this project. The faculty have obtained material property goals from Lehigh University, worked with Lord Corporation to develop candidate materials to meet those goals, investigated various manufacturing processes with Lord Corporation, and coordinated communication between Lehigh University and Lord Corporation.

Lehigh has provided the part-time resources of a senior faculty and a graduate student to this project. Lehigh has a long history in the area of structural seismic control and has issued many publications in this area. See References for a partial listing. For this project, they have provided computer resources for analytical modeling, determined material property goals from the modeling, and provided basic configuration needs for the dampers.

Lord Corporation provided technical support including a part time intern during year one, access to key technical personnel including materials and product engineers, and access to historical files related to seismic devices. The patent office of Lord Corporation was used extensively for researching patents and utilizing patent search engines. Finally, several candidate materials were selected and tested by Lord Corporation. The Penn State team and intern utilized Lord Corporation facilities throughout year one working out of the Product Development Group.

Table 1 outlines the strategic actions, collaboration and target timing of the project.

Year One Accomplishments

Year one accomplishments include identification of material property goals, development or selection and testing of elastomers to meet those material property goals, development of a preliminary design concept, the background research of past and existing relevant technologies, and research into past and existing marketing successes and failures of seismic devices.

Material Property Goals

Based on analytical dynamic modeling of energy dissipation devices, Lehigh University has provided elastomeric material property goals to Penn State Erie which have been forwarded to Lord Corporation. Ideally, the material should have a loss factor of approximately 0.6 although a loss factor of approximately 0.4 would be adequate. To be practical for design purposes, the material should have a complex shear modulus of approximately 150 psi.

Just as importantly, the material properties of loss factor and shear modulus must not vary significantly when subjected to the expected ranges of frequency, strain amplitude, and temperature. The material must be insensitive to these varying conditions to overcome the known disadvantages of current elastomeric dampers. These material properties also must not change over time. Details of these and other material property goals are provided in the Appendix.

Status of Accomplishing Material Property Goals

Through year one, four elastomers, two each of two families, were thoroughly tested by Lord Corporation under the coordination of Penn State Erie for sensitivity to material property changes over the expected frequency, strain, and temperature ranges. Tests to date have been conducted on laboratory specimens under laboratory-controlled conditions.

Figures 1 and 2 show sample results for one elastomer from the organic family and figures 3 and 4 show sample results for one elastomer from the silicone family, all at 68°F (20°C). The detailed results for only one elastomer from each family are presented here, accompanied by generalized observations for each elastomer family. Complete details of the results will be published elsewhere upon completion of the project.

Figures 1 and 3 show that the target complex shear modulus of 150 psi is possible with both the silicone and organic elastomer families. Figures 2 and 4 show that the target loss coefficient will be possible with this organic family but not with this silicone family. Figure 3 shows that, for the silicone elastomer, the complex shear modulus is virtually independent of frequency although it varies somewhat with strain amplitude. Figure 1 shows that the complex shear modulus of the organic elastomer varies slightly with frequency but varies significantly with strain amplitude. The relationships and trends for loss coefficients are observed to be the same as those for complex shear modulus. Although the results are not shown here, additional tests indicate that the properties of the silicone are less sensitive to changes in temperature than those of the organic elastomer.

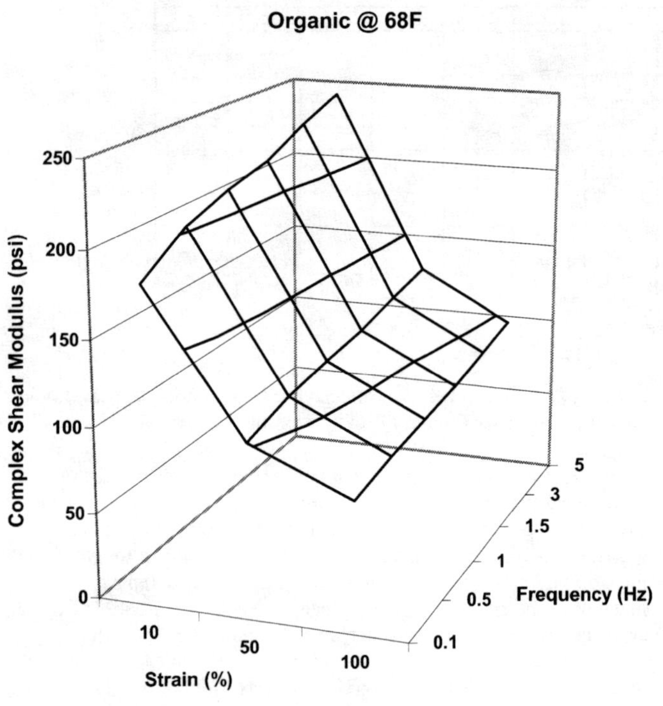

Figure 1

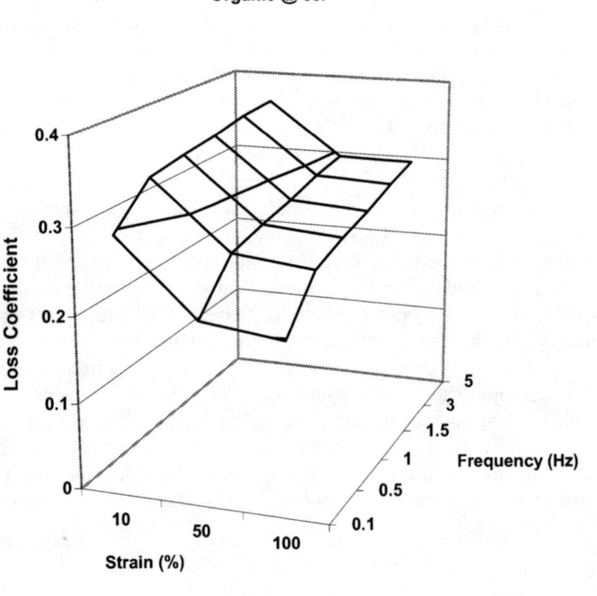

Figure 2

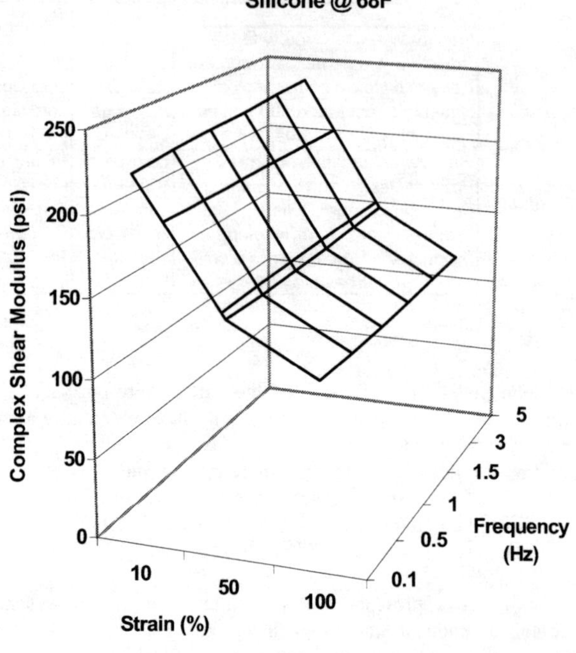

Figure 3

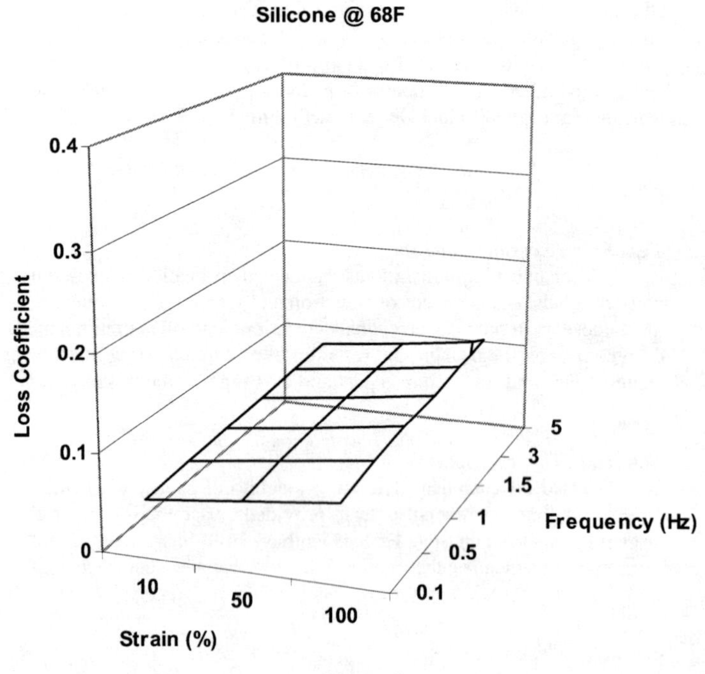

Figure 4

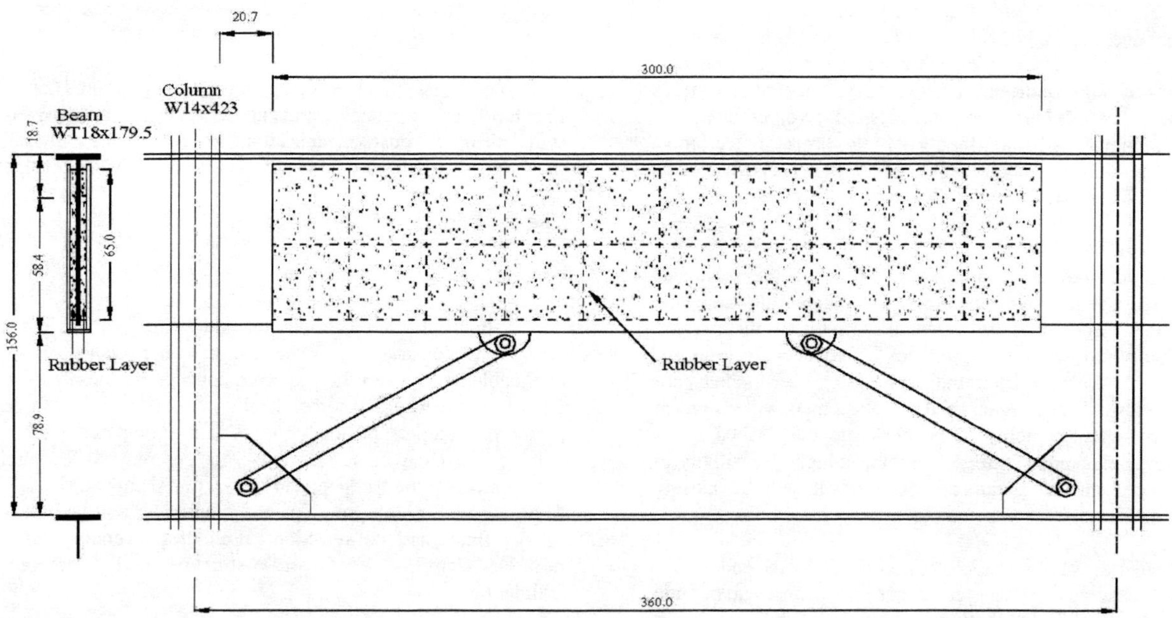

Unit = inches
Thickness of Damper = 2.67 inches (2"(story drift)/2.67 = 0.75%)
Damper Area = 2 layers x 65 in (height) x 300 in (width) = 39000 sq. inches

Figure 5

This silicone family, although not a highly damped, was tested because it is a low-cost silicone which may offer competitive advantages. This organic family is a relatively highly damped organic elastomer and is less costly than any silicone elastomer. The family of organic elastomers is being considered and tested because of its known, large cost advantage over all silicone elastomers. Other potential advantages of organic elastomers may include better bondability, higher tensile strength, and better fatigue properties.

Although material tests are being conducted on a wide range of strain, frequency, and temperature conditions, the condition of 100% strain, 0.5 Hz, and 68°F has become the baseline for the material properties to be used in damper design. Other conditions are tested to evaluate the sensitivity of properties from this baseline condition. All material testing results to date have been forwarded to Lehigh. Lehigh is optimistic that material goals can be reached so that the elastomer can function as desired in structural dampers.

Projected Concept and Design

Lehigh University has developed preliminary design concepts that should physically satisfy the damper properties used in the analytical structural models. Design sketches have been provided to Penn State Erie by Lehigh and can be seen in Figure 5. Generally, the damper design concept and configuration consists of two layers of elastomer bonded between three steel plates. The two outer steel plates are connected, outside of the elastomer configuration, and act as one member. When there is relative motion between the inner steel plate and the outer steel plates, the elastomer is deformed in shear, resulting in energy dissipation.

The inner steel plate may be the vertical web section of an existing tee-shaped beam in the structure. Bracketry and linkages are then used to connect the outer plates to a lower level in the structure. The motion of the damper, which results in energy dissipation, will predominantly be the lateral motion within the structure from one level to the next.

The sections of elastomer used to dissipate energy from one level to the next are large and may by up to 300 inches long by 65 inches high. Each elastomer section will be approximately 2 inches thick. These sections are so large that, due to current tool size limitations, segmented construction will probably be necessary. Penn State Erie has started discussions with manufacturing experts at Lord Corporation to select or develop the appropriate manufacturing processes for metal preparation, adhesive application, bonding, curing (vulcanization), etc. to obtain the desired configuration.

The research of past and existing relevant technologies is summarized in the Background section of this paper. The research into past and existing marketing successes and failures of seismic damper products has led to the proposed extension of the project, Commercial Feasibility Study, described in the goals for year two.

Year Two Goals

Major goals for year two are to finalize development of material and provide Lehigh with material test data; design, develop and prototype full scale dampers; and investigate commercial markets for potential elastomeric dampers. Details of each goal are summarized below.

Material Development Plans

The test results for both the silicone and organic elastomer families were encouraging but they do indicate that further material development is necessary. The silicone elastomers, although insensitive to varying conditions of frequency, temperature, strain amplitude, and time, did not have the desired level of damping. Consequently, another silicone family of elastomers is currently being tested. This second silicone family of elastomers is expected to have the desired level of damping but is known to be more costly.

The organic elastomer family did have damping that approached the desired level but the modulus of the particular elastomers tested was too low to be practical in damper design. This was due to the higher, and unknown, sensitivity of the organic elastomer family to various strain and frequency conditions. Consequently, elastomers of the same family but higher modulus are currently being tested. The higher modulus elastomers will also have higher loss coefficient which will be closer to the desired level. Since the organic elastomer family did exhibit a higher sensitivity to strain, frequency, and temperature conditions, it will be closely monitored in future tests.

All elastomer tests above should be completed by the end of April 2003. A candidate silicone elastomer and a candidate organic elastomer will be identified at that time to proceed with to the design phase.

Damper Design and Construction

Design and configuration ideas will be subjected to a form of the Advanced Product Quality Planning (APQP) process as outlined by the Automotive Industry Action Group (AIAG) to satisfy QS9000 requirements. Lord Corporation utilizes this process, or variations of it, in their Industrial Products Division which is the likely assignment of structural dampers within the company.

The three major components of the APQP process that will be coordinated by Penn State Erie for this project are 1) Product Design and Development, 2) Process Design and Development, and 3) Product and Process Validation. Product Design and Development is being done with the ongoing collaboration of Lehigh University, Lord Corporation, and Penn State Erie. Process Design and Development, which has started, will be done primarily by Lord Corporation and Penn State Erie. Product and Process Validation will be done primarily by Lehigh University and Penn State Erie.

Some tools established by Lord Corporation for Product Design and Development are Design Failure Mode and Effects Analysis (DFMEA), Design for Manufacturability and Assembly (DFMA), Concept Design Reviews, Final Product Reviews, etc. These tools, if used, will involve applicable personnel from all three organizations and will provide verification that the designs being developed will meet the overall technical goals of the project. In addition to verifying that all technical goals are met, this APQP process will provide documented evidence of design evolution, changes, and reasons.

Penn State Erie is currently working with Lord Corporation on the Process Design and Development necessary to build damper prototypes. Items being investigated by Penn State Erie and Lord Corporation are 1) the means of accomplishing the segmented construction with current tool size limitations, 2) calendaring vs. molding processes, 3) bonding processes of calendared layers of elastomer and of elastomer to metal, and 4) the need and configuration for cores to provide the necessary heat transfer during curing.

At least three molding processes (compression, transfer, injection) are available and at least three bonding processes (vulcanization, post-vulcanization, cold) are available. These investigations will be conducted with both the Mechanical Products and the Chemical Products Divisions of Lord Corporation. The Chemical Products Division has expertise in the bonding processes and the Mechanical Products Division has expertise in both the bonding and molding processes.

A goal of Lord Corporation and Penn State Erie will be to design and construct prototype dampers as close as possible, upon applicable reviews, to the projected concept and design provided by Lehigh University. Another goal is to have the configuration of prototype dampers be the same as that anticipated for production dampers. Actual manufacturing processes may differ significantly between prototype dampers and production dampers. Final selection of processes for both prototype and production will be based, in part, on the final material selection. Lehigh University will later test prototype dampers in a full-scale structure for Product and Process Validation.

Commercial Feasibility Study

A third and important goal for year two is to validate the commercial potential for elastomeric dampers. Lord Corporation has requested that the research project have a minor extension of financial and personnel resources to include a business development and marketing element in light of past difficulties entering these markets. This work will leverage existing marketing and business studies already completed by Lord. One such study, completed by an outside consulting firm, provides a detailed review of every major damper manufacturer worldwide (all types – viscous, friction, rotary). This study encompasses all major markets including aerospace, automotive, industrial, railroad and seismic. Nevertheless, several questions and issues related to elastomeric seismic dampers still remain. Some of these questions have been identified in initial meetings with the Penn State Erie School of Business, Lord Corporation and various discussions with Lehigh. A brief list follows:

1. How do seismic engineers specify building dampers? What technical specifications or codes are used?
2. Who ultimately drives the decision for use of dampers? What are the potential sales channels for elastomeric seismic products? Who should the products be marketed to? (seismic engineers, architects, civic authorities, building owners, insurers, other)
3. What size ranges are needed to capture all or key portions of the building market? What portion(s) of the building make the most sense to target with elastomeric dampers?
4. Does the concept of a simple, relatively low-cost damper (designed and implemented within conventional building design and construction practice) have any value to the market?
5. Competing technologies have already been identified but what competitive advantage, other than possibly cost, does an elastomeric damper have over existing products such as fluid dampers?

6. How much is the market willing to pay for successful dampers? What are the anticipated development and testing costs for elastomeric dampers? How long is any payback period?

CONCLUSION

The collaborative efforts of this research project show how the applied research capabilities of a mechanical engineering technology program can complement the pure research capabilities of a leading institution in their field. Further, the collaborative efforts of this research project show how industry can be linked with research to develop and advance technology. The industry influence on this project will have a positive impact on bringing a research technological concept to market.

REFERENCES

[1] Higgins, C. C., Kasai, K. 1997 "Real-Time Testing and Analysis of a Full-Scale Viscoelastically Damped Steel Frame", ATLSS Report No. 97-14, Department of Civil and Environmental Engineering, Lehigh University.

[2] Lee, K.-S. & Sause, R. & Ricles, J. & Lu, L.-W. & Ab-Malek, K. "Seismic Behavior of Steel MRF's with UHDNR structural Dampers"

[3] Fan, C.-P. 1998 "Seismic Analysis, Behavior, and Retrofit of Non-Ductile Reinforced Concrete Frame Buildings with Viscoelastic Dampers", Ph.D. Dissertation, Department of Civil and Environmental Engineering, Lehigh University.

[4] Kasai, K. & Munshi, J.A. & Lai, M. & Maison, B.F. 1993. "Viscoelastic Damper Hysteretic Model: Theory, Experimental, and Application", Seminar on Seismic Isolation, Passive Energy Dissipation, and Active Control, ATC-17-1, San Francisco.

[5] Lee, K.-S. 2003. "Seismic Behavior of Structures with Structural Dampers Made from Ultra High Damping Natural Rubber", Ph.D. Dissertation, Department of Civil and Environmental Engineering, Lehigh University.

[6] Sause, R. & Lee, K.-S. & Ricles, J. & Ab-Malek, K. & Lu, L.-W. 2001 "Non-linear Hysteresis Models for Ultra-high Damping NR Structural Dampers", Journal of Rubber Research Volume 4

[7] Sause, R., Hemmingway, G., and Kasai, K. (1994). "Simplified Seismic Response Analysis of Viscoelastic-Damped Frame Structures", *5NCEE*, Chicago, IL, 834-848.

APPENDIX

Material Property Goals

1. The material should have a loss factor of approximately 0.6. A higher loss factor is desirable if there is no negative effect on frequency or strain sensitivity.
2. The material should have a shear modulus of approximately 150 psi at 100% strain, 0.5 Hz, and room temperature.
3. Material properties, particularly shear modulus and loss factor, should not vary by more than 20% in the temperature range of 10°C to 40°C.
4. Material properties, particularly shear modulus and loss factor, should not vary by more than 20% in the frequency range of 0.1 Hz to 5.0 Hz.
5. Material properties, particularly shear modulus and loss factor, should not vary by more than 20% in the cyclic strain range of 10% to 200%.
6. Material must withstand several cycles of 200% shear strain in service without damage or change of properties.
7. Material must withstand several cycles of 500 psi shear stress without damage or change of properties.
8. Material bond must withstand several cycles of 500 psi shear stress without damage.
9. Material properties, particularly shear modulus and loss factor, must not change by more than 20% in 25 years of sitting idle.
10. Material must withstand only minimal environmental exposure.

Proceedings of IMECE'03
2003 ASME International Mechanical Engineering Congress
Washington, D.C., November 15–21, 2003

IMECE2003-41111

A GENETIC ALGORITHM APPROACH TO WELD PATTERN OPTIMIZATION IN SHEET METAL ASSEMBLY

Gene Y. Liao
Wayne State University
Detroit, Michigan 48202
Email: liao@et.eng.wayne.edu

ABSTRACT

In sheet metal assembly process, welding operation joins two or more sheet metal parts together. Since sheet metals are subject to dimensional variation resulted from manufacturing randomness, gap may be generated at each weld pair prior to welding. These gaps are forced to close during a welding operation and accordingly undesirable structural deformation results. Optimizing the welding pattern (the number and locations of weld pairs) of an assembly process was proven to significantly improve the quality of final assembly. This paper presents a Genetic Algorithm (GA)-based optimization method to automatically search for the optimal weld pattern so that the assembly deformation is minimized. Application result of a real industrial part demonstrated that the proposed algorithm effectively achieve the objective.

NOMENCLATURE

f value of objective function
n number of output/critical points for objective function
w_i weight factor for deformation at *i*th point
δ_i mean deformation at *i*th point

1 INTRODUCTION

In sheet metal assembly process, welding is an important operation which joins two or more sheet metal parts together. The weld spots also increase the internal stress in these flexible sheet metal parts. The pressure and heat exerted by the spot welding equipment tend to deform the sheet metal to form the assembly joint. Additionally, since each sheet metal is subject to dimensional variation caused by manufacturing randomness, gap may be generated at each weld pair prior to welding. These gaps are forced to close during a welding operation and accordingly undesirable structural deformation results. Therefore, welding pattern (the number of weld pairs and their locations) has great impact on the final build quality. In current practice, however, there are no straight rules to be applied. It is highly desirable to develop a systematical tool to meet this need. This paper intends to address this issue.

Welding induced distortion is mainly dominated by the number and locations of weld joints as well as welding sequence. A number of approaches have been developed to optimize the designs of sheet-metal welding operations in the areas of weld pattern and welding sequence [1-5]. These approaches calculate residual stress and deformation of an elastic workpiece welding assembly, and have a wide range of applications. Although these approaches had their successes on achieving respective goals for a successful welding configuration, it is noted the optimal selection of the positions of welding points was based on a stationary set of welding condition. That is, the number of the weld points was fixed during optimization. The constrained number of weld points in optimizing their positions does not ensure that resulting solution is effective or optimal. This is especially true of the assembly large sheet metal panels.

Dimensional variation of each part is resulted from its preceding manufacturing process, such as forming operation. The dimensional variation and deformation of individual part will cause final assembly variation. The objective of this paper is to develop a Genetic Algorithm (GA) approach to automatically select the optimal numbers of weld points as well as their optimal positions in sheet metal assembly, such that the final welded assembly deformation due to individual part dimensional variation is minimized.

Imitating the principles of natural evolution, GA is an effective way to find the optimal design involving discrete variables. In GA, a set of design alternatives that form a population in a given generation, is allowed to reproduce and cross breeds among themselves to create a new set of designs in the next generation, with bias favoring the most fit members in the population. Combination of the most desirable characteristics of mating members of the population results in new design alternatives that are more fit than their parents. The GA has been proven to be a useful technique in solving constrained engineering problems and optimization problems in the area of welding design [1, 4, 5].

2 GA-BASED WELD PATTRN OPTIMIZATION

The weld pattern optimization problem is defined as: to select the number and location of the weld spots so that the objective function, which will be defined in section 2.2, is minimized.

2.1 Design Parameters and Pools

For a GA-based optimization method, design pools that contain all potential design candidates (welding locations) need to be pre-defined. These pre-defined candidates should be feasible and are selected by users according to their engineering knowledge and experience. Due to geometrical constraints, weld points could not be placed in some areas of the part, such as pierced holes in the part. The GA will then search for the best design candidates automatically from the pools to reach the design objective. It is therefore essential to select all the possible points in the physical part and collect them into the weld pool. An example is shown in Fig. 1, where weld pool consisting of 8 members (1, 2, 3, 4, 5, 6, 7, 8) is defined in part 1. Since welds appear in pairs and therefore there are also 8 corresponding points in part 2 (not shown).

Another parameter that needs to be defined in weld pattern optimization is the maximum number of welds to be used in the welding operation. GA will search automatically for the best number and location of welds to meet the design objective. For example, (2, 3, 6, 8) in Fig. 1 may be obtained after optimization with the input of five as the maximum number of welds. In this case only four welds at locations 2, 3, 6 and 8 are needed for minimal assembly deformation. With the number of welds reduced from five to four, the welding cycle time and cost are possibly reduced.

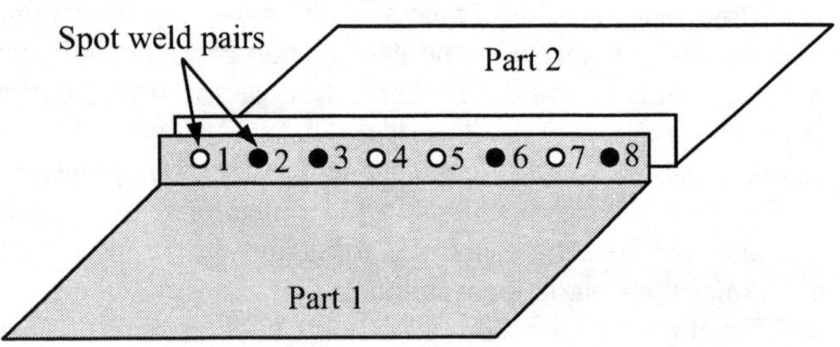

Welding Pairs Design Pool 1, 2, 3, 4, 5, 6, 7, 8

Figure 1. Weld Pool Example

2.2 Design Objective Function

The objective function can be either deterministic or random. For the deterministic case a typical objective function is

$$f = \sum_{i=1}^{n} w_i \delta_i^2 \tag{1}$$

where δ_i is the deformation at ith point, w_i is the corresponding user-defined weight factor at ith point, and n is the number of interest points.

2.3 String Representation

There are questions facing the application of the GA in this type of problem defined previously. For example, how do we treat a variable number of welds and generate new designs in GA mutation? In this section a new GA is proposed to resolve these issues. Additionally, the design parameters (the physical system of weld points) need to be encoded into a GA string.

A mutation operator is used to generate new designs. It exchanges one or more randomly selected members (weld spots) in a GA string with one or more randomly selected members in the corresponding design pool. The GA string representation for the shaded welds shown in Fig. 1 is (2, 3, **6**, 8) and the corresponding pool is (1, **4**, 5, 7). Note that the pool only contains the members that are not selected for a corresponding design. If "6" and "4" are randomly selected from GA string and design pool, respectively, the new design becomes (2, 3, **4**, 8) after mutation operation. The four permutation operators presented in [3] can also be applied here to optimize the welding sequence.

To treat the variable number of welds in GA search, dummy members, 0, are added into the design pool for selection. If a dummy member is selected to replace a string member during mutation operation, then this newly generated design will have one less number of welds than the original design. All of the dummy welds will be automatically ignored by the algorithm.

This study applies finite element analysis (FEA) to model the workpiece and hence weld locations are restricted on those nodal points [6]. The dimensional size of element influences the candidates of sub-pools and solution. The smaller dimensional size of the element, the larger number of nodal points and consequently the larger number of candidates in sub-pools can be generated. However, a huge number of nodal points or elements might need longer computation time. For the purpose of modelling accuracy and effective computation, the maximum length size of each element is limited to 25 mm, which is less than a regular size of weldment.

3 OPTIMIZATION PROCEDURE

The optimization procedure is depicted in Fig. 2. The number of population (*np*) in GA strings is generated by random selection of members from the weld pool. These *np* designs are evaluated by the objective function. GA operations, which reproduce and mutate with dummy members, are then applied to generate the strings for the next generation. This process is repeated until objective function is converged or the maximum iteration number is reached.

4 APPLICATION EXAMPLE

Figure 3 shows the finite element model of two typical vehicle sheet metal parts: dash and dash-front-end panels. The two sheet metal parts are assembled through the welding operation. The clamping schemes for the welding operation are also shown in Fig. 3. The key measure for assembly quality is the deformations at interest points, which are shown in Figure 4. Deviations of the parts are the sources of deformation for the study. Since this is a deterministic study, the objective function shown in Eq. (1) is used.

The weld pool, which consists of 31 weld points, is also shown in Fig. 3. Note that these points are all on dash panel. Since welds appear in pairs for welding operation, there are also 31 corresponding points on dash front-end panel (not shown in Fig. 3). These 31 points are the design candidates. An initial weld pattern with 14 weld pairs is shown in Fig. 5. It will be used for comparison with the optimal result.

A population size of 15 was used to perform the optimization. In this case, weld number is fixed to be 14. This implies no dummy members were added to the design pool for mutation. It converged after 46 iterations of GA search. The comparisons between the initial and optimal weld designs are listed in Table 1. The objective function is significantly reduced from 5.922mm to 0.025mm, a 99% reduction. Figure 6 shows the optimal weld locations solved by the algorithm. The optimization history is plotted in Figure 7.

To verify the simulated objective functions, an in-process coordinate measuring machine will be used to measure the deflection at specified critical points. The dimensional variation at the critical points will be measured right after the sheet-metal forming process. Six workpieces sampled from the same production line will be measured to form the objective function. This validation process is on the planning.

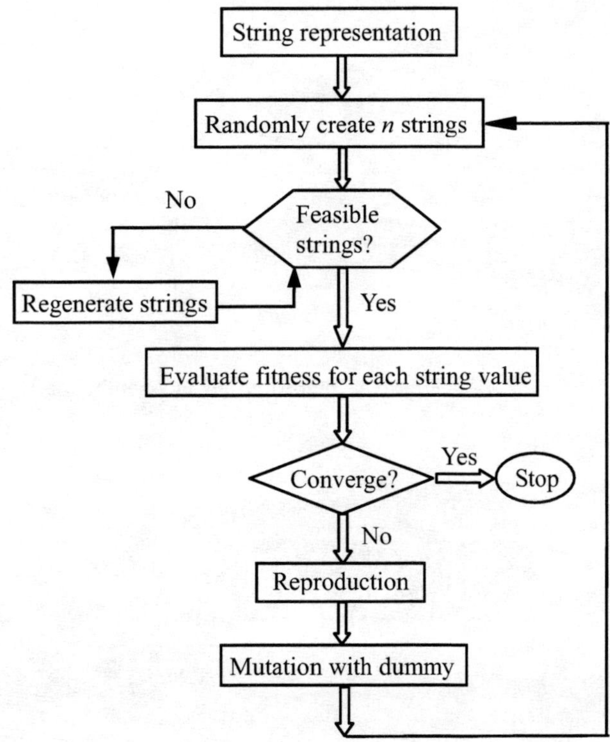

Figure 2. Flowchart of GA search procedure

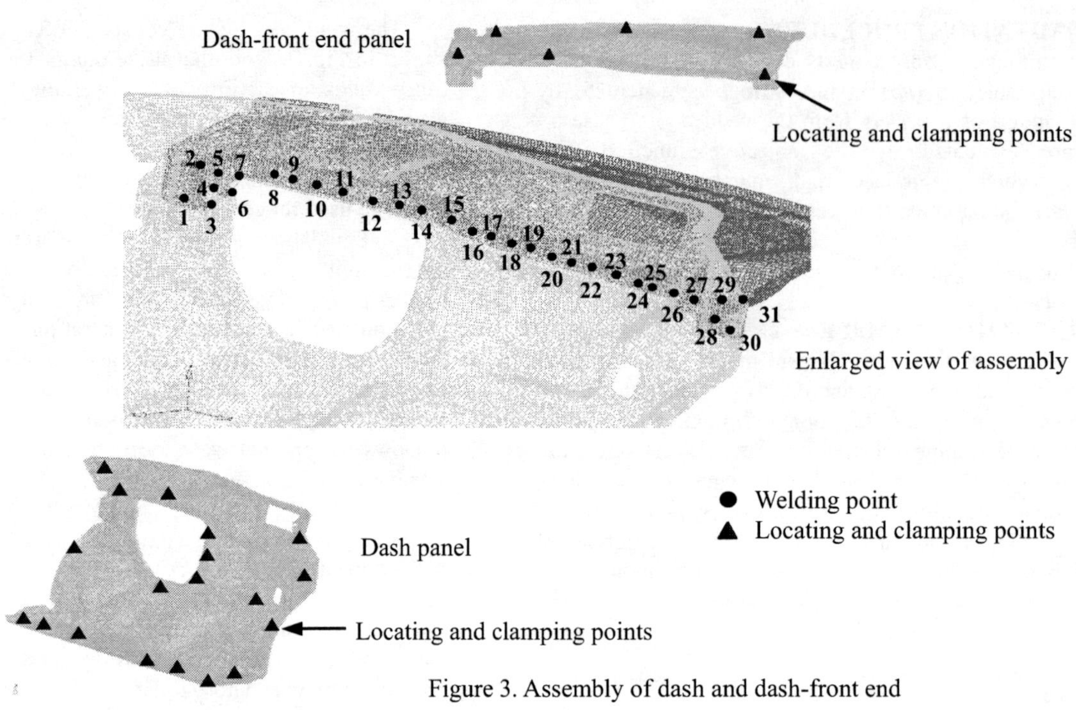

Figure 3. Assembly of dash and dash-front end

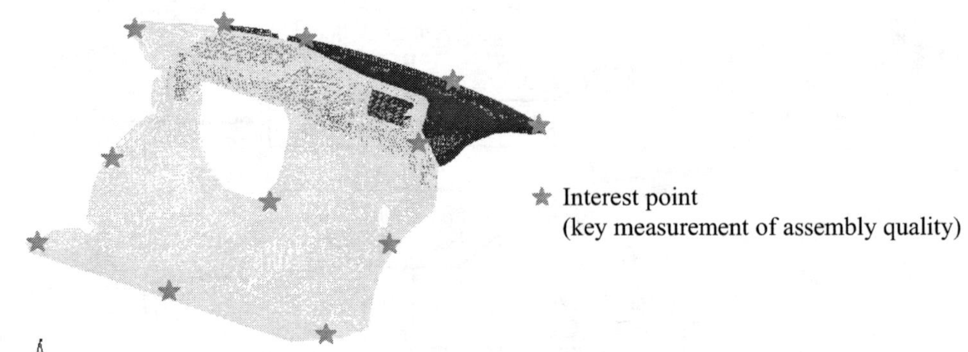

Figure 4. Interest points in assembly part

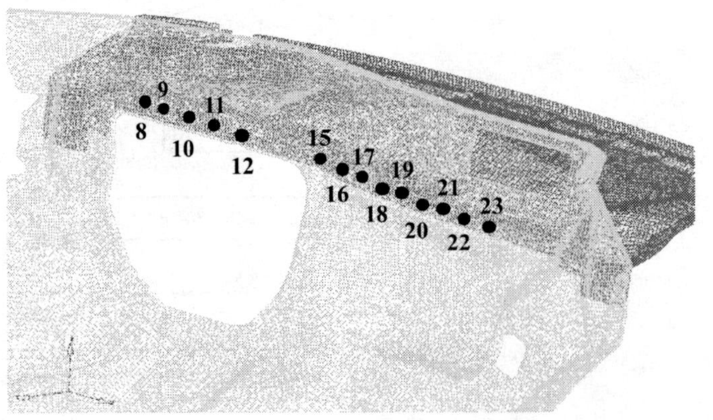

Figure 5. Initial weld pattern with 14 welds

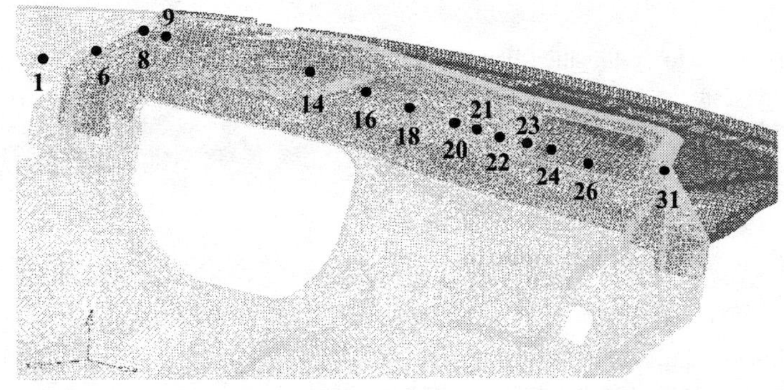

Figure 6. Optimal weld pattern with 14 fixed welds

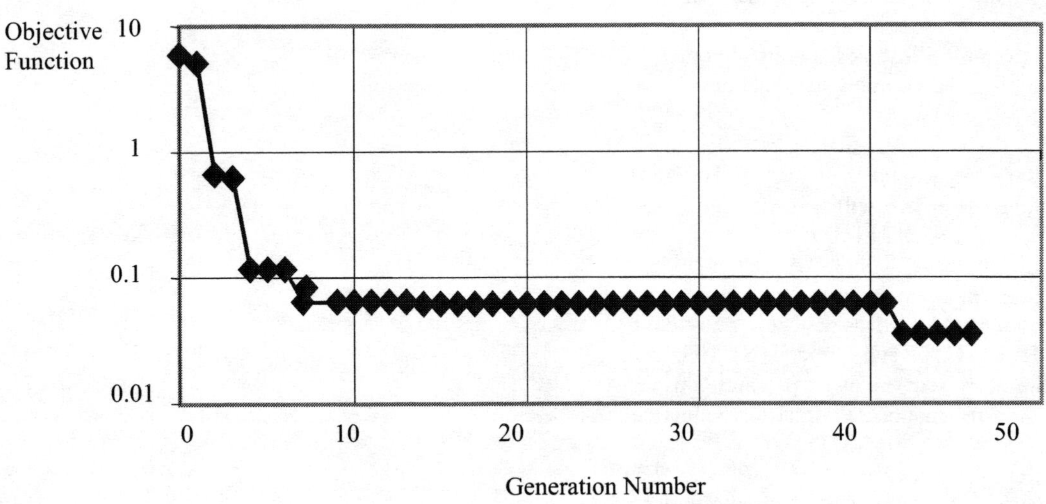

Figure 7. Optimization history plot with 14 fixed welds

Table 1. Initial and optimal designs

Initial Pairs	Optimal Pairs
8,9,10,11,12,15, 16,17,18,19,20,21,22,23	1,6,8,9,14,16,18, 20,21,22,23,24,26,31
Objective function 5.922 mm	Objective function 0.025 mm

5 CONCLUSIONS

A GA-based method is presented to systematically search for the optimal numbers of weld spots as well as their optimal positions in sheet-metal assembly process, such that the workpiece deformation due to assembly variation is minimized. The objective function is formulated in deterministic or random case. The number of welds can be fixed or variable. With the variable weld number optimization, welding cost may be significantly reduced due to less number of welds.

To treat the variable number of welding points during GA search, dummy members are added into the welding points pool for selection. If a dummy member is selected to replace a string member during mutation, then this newly generated design will have one less number of welds than the original design. This proposed treatment overcomes a problem of string representation for the variable welding points in the GA mutation.

This application example illustrates the effectiveness of the algorithm in minimizing the assembly deformation.

REFERENCES

[1] Kadivar, M. H., Jafarpur, K. and Baradaran, G. H., 2000, "Optimizing welding sequence with genetic algorithm," Computational Mechanics, 26, 514-519.

[2] Hsieh, C. C. and Oh, K. P., 1997, "A framework for modeling variation in vehicle assembly processes," International Journal of Vehicle Design, 18(5), 466-473.

[3] Huang, M. W., Hsieh, C. C. and Arora, J. S., 1997, "A genetic algorithm for sequencing type problems in engineering design," International Journal for Numerical Methods in Engineering, 40, 3015-3115.

[4] Shiu, B. W., Shi, J. and Tse, K. H., 2000, "The dimensional quality of sheet metal assembly with welding-induced internal stress," IMechE, Journal of Automobile Engineering, 214, 693-704.

[5] Chae, S. W., Kwon, K. Y. and Lee, T. S., 2002, "An optimal design for spot welding locations," Finite Element in Analysis and Design, 38, 277-294.

[6] *MSC/NASTRAN version 70.7.2 manual*, 2001, The MacNeal-Schwendler Corp., Los Angeles, CA.

Proceedings of IMECE'03
2003 ASME International Mechanical Engineering Congress
Washington, D.C., November 15–21, 2003

IMECE2003-41332

Computer-Aided Power Requirements

Edward M. Vavrek
Purdue University North Central

Introduction

The power requirement of machinery is a necessary consideration for the sizing and selection of machine elements in the design process. The decision process of designing gears, belt drives, clutches, brakes, and bearings are designed and sized based on the power requirement of the application. I have written a software program that calculates the power requirements for three different design applications. This software program simplifies and streamlines the design and selection process by allowing the student to calculate the horsepower and torque requirement.
The program is written in Visual Basic, which is an event driven program. The student can go through the program inputting values into text boxes and initiating events by clicking on command buttons. An important feature of this program is its ability to insert pictures, tables, graphs, drawings, and figures to enhance the programs functionality. The program is visual and interactive which helps guide the user through the design process. While the user should have some basic knowledge of torques, speed, and horsepower, the program assists the user through the design process. Most inputs are taken from graphs, charts, or tables so the user knows where and how the information is obtained and used. Outputs are shown with their appropriated figures, drawings, or calculations to assist the student in understanding what the output means. The software has the advantage of calculating minor changes to a problem easily and quickly. This software program is used as a tool to help students learn the design process. This paper will discuss three different design requirement programs and outcome of using the software in the course

Drive Analysis Program

The program starts with a main menu shown in figure 1. The user has the option of selecting three different drive arrangements to analyze. The three drive arrangements are a rotating roller, conveyor drive, and a leadscrew drive. Inertias, torques, and horsepower requirements are calculated for the these drive arrangements.

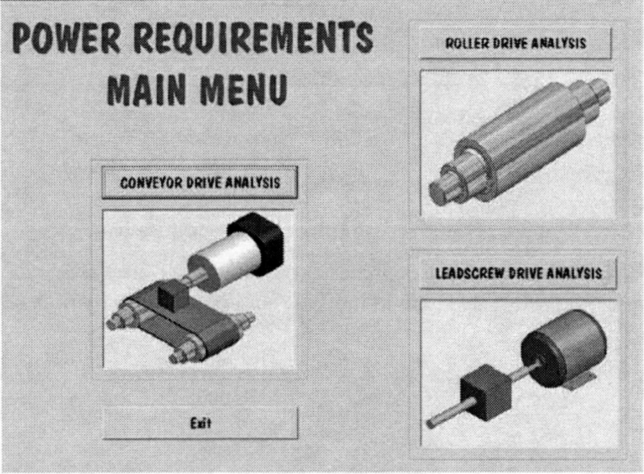

Figure 1. Power RequirementsMain Menu.

Nomenclature

Wt = Weight, [lbs]
A = Cross-sectional area, [in^2]
L = Length of roller, [in]
ρ = density of material, [lbs/in^3]
I = inertia of body, [$lb\text{-}in\text{-}s^2$]
v_o = initial velocity, [in/s]
v_f = final velocity, [in/s]
$v_{surface}$ = surface velocity of roller, [in/s]
D = diameter of roller, [in]
R = radius of roller, [in]
n = angular velocity of roller, [rpm]
ω = angular velocity [rad/s]

t = time, [sec]
a = linear acceleration, [in/s^2]
α = angular acceleration, [rad/ s^2]
e = leadscrew efficiency
μ = leadscrew coefficient of friction
p = leadscrew pitch, [threads/in]
g = gravitational constant, [32.2 ft/s^2]
N = gear ratio
T = torque, [lb-in]
HP = horsepower, [hp]

Roller Drive Analysis Program

Selecting the "Roller Drive Analysis" command button, will initiate the roller drive analysis program. Figure 2 has the user enter the dimensional specifications of a solid roller. The user has the option to select a solid roller or a hollow roller. If the user selects the hollow roll, figure 2 will change to allow the user to enter the dimensions of a hollow roller.

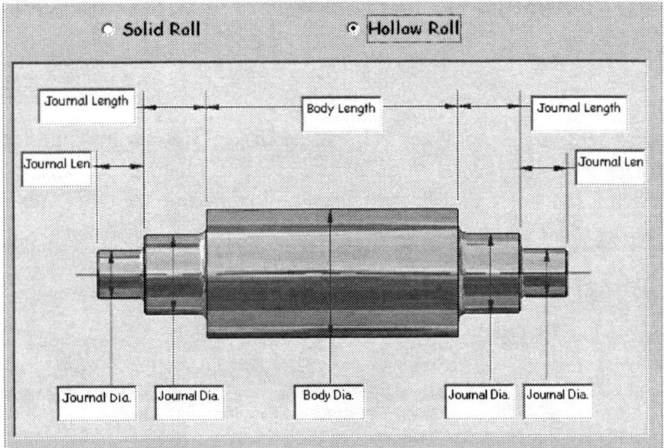

Figure 2. Enter Roller Dimensional Specifications.

The user then selects the material option of the roller. The user can select steel, aluminum, or can input the density of the material.

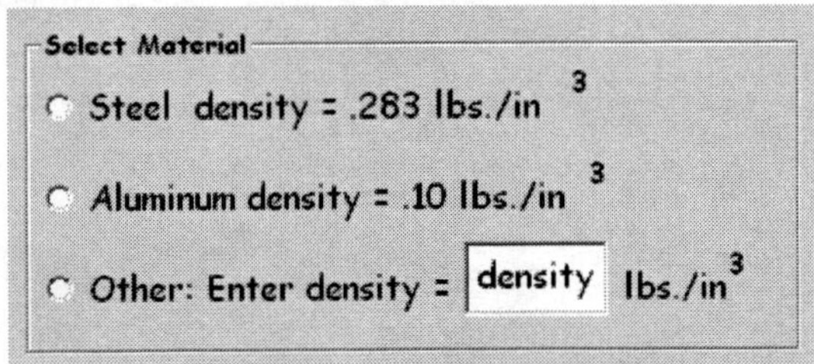

Figure 3. Material Option of the Roller.

From the roller dimensions and the density of the material, the weight of the roller is calculated in the program. The program calculates the weight of the roller using equations 1, 2 and 3.

$$Wt_{BODY} = A_{BODY} \cdot L_{BODY} \cdot \rho_{BODY} \tag{1}$$

$$Wt_{JOURNAL} = A_{JOURNAL} \cdot L_{JOURNAL} \cdot \rho_{JOURNAL} \tag{2}$$

$$Wt_{TOTAL} = Wt_{BODY} + Wt_{JOURNAL} \tag{3}$$

The area of the roller depends on whether it is a hollow or solid. The program can then calculate the mass moment of inertia of the roller based on equations 4, 5, and 6.

$$I_{BODY} = \frac{Wt_{BODY}}{2 \cdot g} \cdot R_{BODY}^2 \tag{4}$$

$$I_{JOURNAL} = \frac{Wt_{JOURNAL}}{2 \cdot g} \cdot R_{JOURNAL}^2 \tag{5}$$

$$I_{TOTAL} = I_{BODY} + I_{JOURNAL} \tag{6}$$

The program also calculates the balancing specification of the roller. This is based on weight and speed. This information is used to balance the roller in manufacturing. The program displays these values in the output portion of the program.
The next option the user has is to either enter the surface speed of the roller or the angular speed of the roller shown in figure 4.

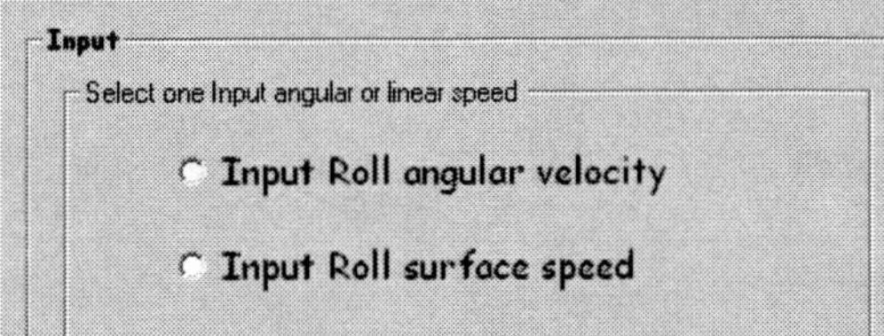

Figure 4. Input Roller Speed.

If the user selects the "Input Roll Surface Speed" option, the program will have the user enter the initial and final surface speeds of the roller shown in figure 5. If the user selects the "Input Roller Angular Velocity" option, the program will have the user enter the initial and final angular velocities of the roller.

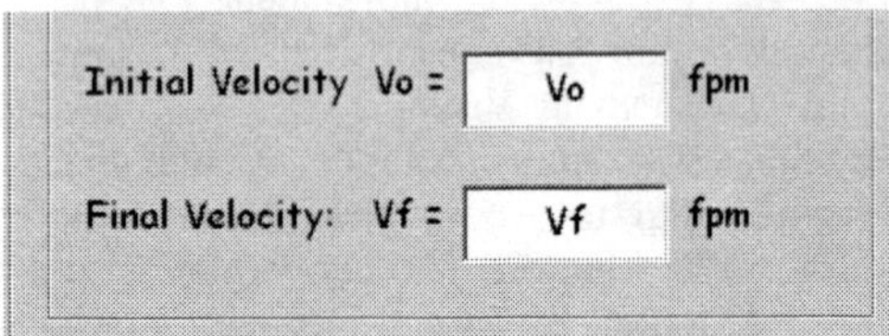

Figure 5. Enter Roller Surface Speed Values.

If the user enters the surface speed of the roller, the program uses the surface speed to calculate the angular speed of the roller. If the user enters the angular speed of the roller, the computer calculates the surface speed of the roller. The computer uses equation 7 or 8 to relate linear and angular speed.

$$v_{SURFACE} = 2 \cdot \pi \cdot \frac{D_{BODY}}{2} \cdot n_{BODY} \tag{7}$$

Equation 7 can be rewritten in terms of the angular velocity of the body shown in equation 8.

$$n_{BODY} = \frac{v_{SURFACE}}{2 \cdot \pi \cdot \frac{D_{BODY}}{2}} \tag{8}$$

The next option has the user enter either the time or acceleration/deceleration of the body shown in figure 6.

Figure 6. Time / Accel Option.

If the user selects the time option, the user is prompted to enter time as shown in figure 7.

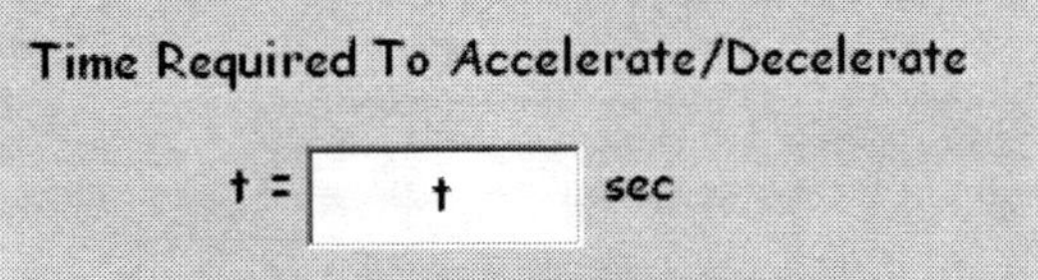

Figure 7. Time Value.

Time is used to calculate the acceleration or deceleration of the roller from the initial speed to the final speed. This time is used in equations 9 and 10 to calculate the angular and linear acceleration of the roller.

$$a_{BODY} = \frac{v_{FINAL} - v_{INITIAL}}{time} \tag{9}$$

$$\alpha = \frac{\omega_{FINAL} - \varpi_{INITIAL}}{time} \tag{10}$$

The user can also enter the linear and angular acceleration of the body if this value is known. The torque[1] required to accelerate/decelerate the roller in a given time is calculated using equation 11.

$$T = I \cdot \alpha \tag{11}$$

Using this torque and the angular speed, the horsepower requirement to accelerate/decelerate the roller is calculated using equation 12.

$$HP = \frac{T \cdot n}{63000} \tag{12}$$

The program calculates the weight, balancing specification, WK^2, torque, inertia, and the horsepower to accelerate/decelerate the roller as shown in figure 8.

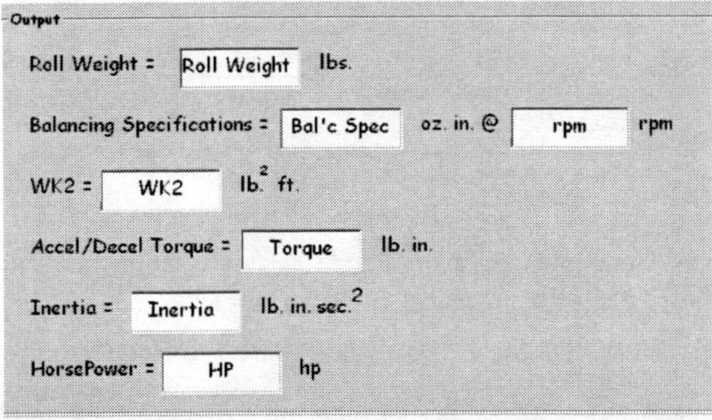

Figure 8. Roller Output.

The output calculated on the roller is used in many applications. The roller weight and balancing specifications are used on the manufacturing drawing of the roller. This gives the machinist the information required to manufacture the roller. The inertia, and horsepower calculated are used to design and size power transmission components to drive the roller. This includes gear drives, belt drives, chain drives, and couplings.

Most drive applications involve the use of a motor. This motor needs to be sized to drive the application. The user has the option to analyze the drive and properly size a motor. This analysis involves calculating the reflected inertia and comparing this inertia with the motor inertia.

Figure 9 has the user input the number of teeth on the belt drive. The drive can be any drive arrangement including a belt drive, gear drive, or chain drive. The ratio is the important factor to calculate the reflected inertia to the motor. The user also enters the motor inertia and motor torque. Since the user is sizing the motor and at this point, he may not know the motor specifications. A starting point will include a guess of the motor. As the user becomes more proficient in drive design and analysis, the more accurate this first guess will be. Also, since designing is an iterative process the user will go through the program more than once.

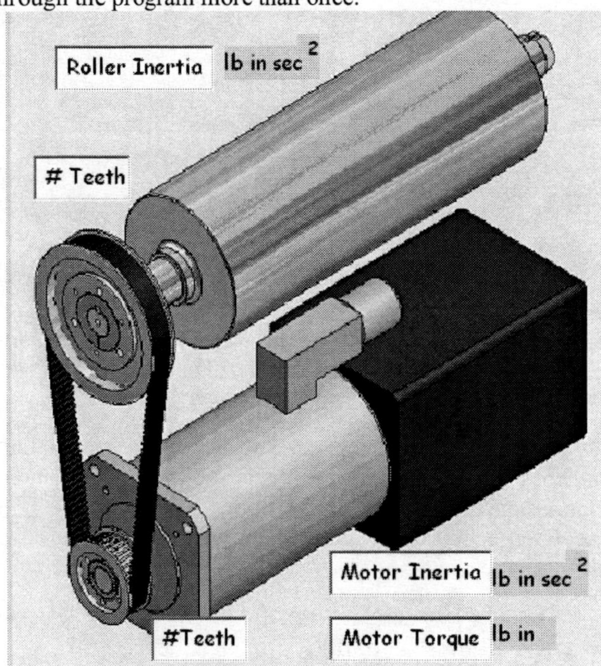

Figure 9. Ratio and Motor Specifications.

The program calculates the reflected inertia at the motor equation 13. This is the load inertia divided by the square of the ratio.

$$I_{reflected} = \frac{I_{load}}{N^2} \tag{13}$$

The total inertia at the motor is the sum of the reflected inertia and the motor inertia equation 14.

$$I_{total} = I_{reflected} + I_{motor} \tag{14}$$

The inertia mismatch between the load and the motor can be calculated. The reflected inertia should not be more than 10 times as much as the motor inertia. The mismatch inertia is given in equation 15.

$$I_{mismatch} = \frac{I_{reflected}}{I_{motor}} \tag{15}$$

The computer program calculates the reflected inertia, total inertia, and the inertia mismatch between the roller and the motor shown in figure 10. These calculations are used in selecting a motor with the correct inertia.

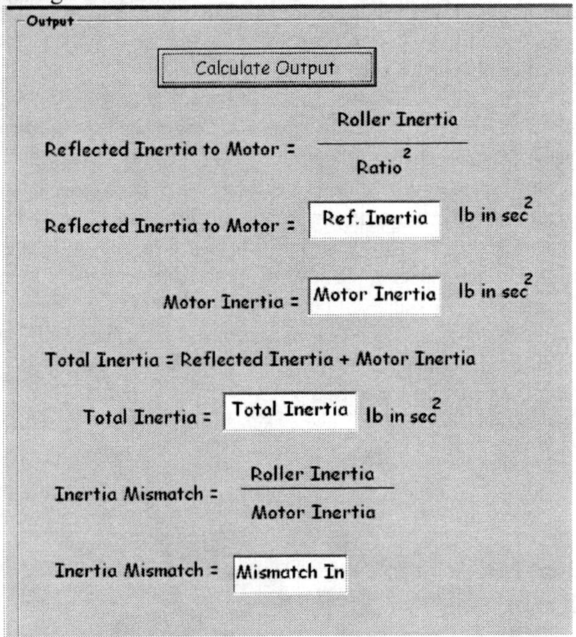

Figure 9. Inertia Output.

The maximum motor speed is compared to the actual running speed of the motor. The torque required to accelerate/decelerate the load is calculated in equation 16.

$$T_{accel} = I_{reflected} \cdot \alpha_{motor} \tag{16}$$

This torque can be compared to the peak torque and continuous torque rating of the motor. The last calculation performed in the program is the power requirements. The horsepower is based on required torque and speed of the system equation 17.

$$HP = \frac{T_{accel} \cdot n_{motor}}{63000} \tag{17}$$

The torque, speed, and power requirements for a roller drive are shown in figure 11.

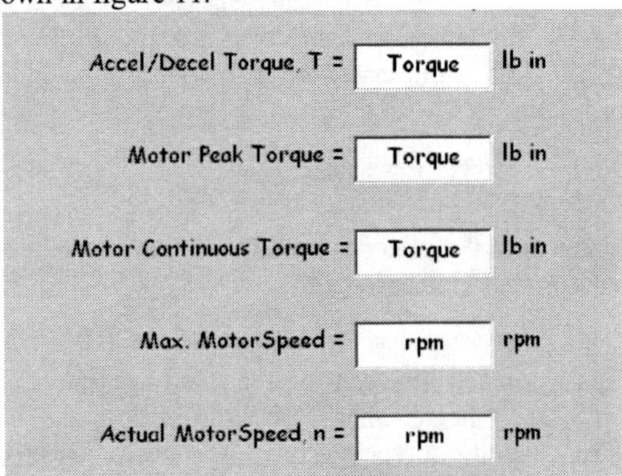

Figure 10. Power Requirements.

Conveyer Drive Analysis Program

The conveyor drive used for analysis consists of two rollers, a belt, a load, a gear reducer, and an electrical motor. The electric motor is sized based on the acceleration needed to get the load to running speed. The user selects the Conveyor Drive Analysis command button shown in figure 12.

Figure 11. Conveyor Drive Analysis.

The user enters the material and size of the roller shown in figure 13.

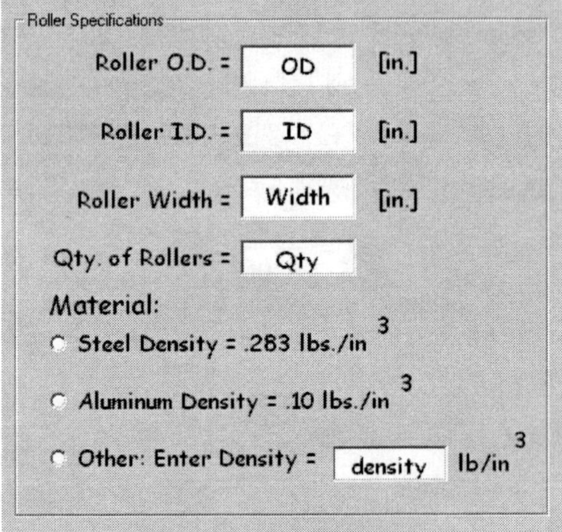

Figure 12. Conveyor Roller Inputs.

The program uses these inputs to calculate the weight and inertia of the roller, using equations 18 and 19.

$$Wt_{ROLLER} = A_{ROLLER} \cdot L_{ROLLER} \cdot \rho_{ROLLER} \tag{18}$$

$$I_{ROLLER} = \frac{Wt_{ROLLER}}{2 \cdot g} \cdot \left(R_{OUTER}^2 - R_{INNER}^2\right) \tag{19}$$

The user enters the speed and weight of the load shown in figure 14.

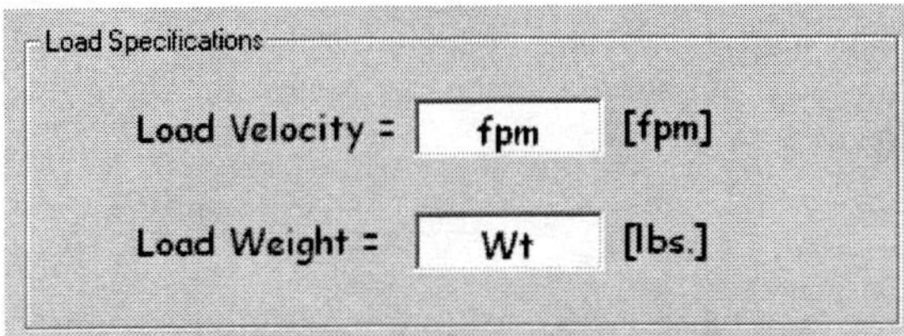

Figure 13. Load Velocity and Weight.

The load velocity is used to calculate the roller speed using equation 19.

$$n_{LOAD} = \frac{v_{LOAD}}{2 \cdot \pi \cdot \frac{D_{ROLLER}}{2}} \qquad (19)$$

The weight of the load is used to calculate the inertia of the load using equation 20.

$$I_{Load} = \frac{WT_{Load}}{g} \cdot a_{Load} \qquad (20)$$

The user has the option of either entering the time or acceleration rate of the load to reach the running speed shown in figure 15.

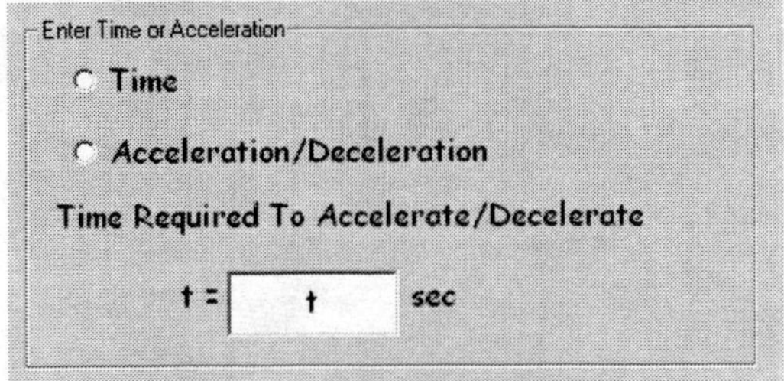

Figure 14. Enter Time or Acceleration Rate.

The user enters the motor inertia and the gear reduction used shown in figure 16. This is a starting point in the design. The required horsepower, torque, and motor inertia will be calculated using this program. The motor size can change therefore changing the motor inertia. The user can also enter the friction torque. This value represents bearing friction the system has to overcome.

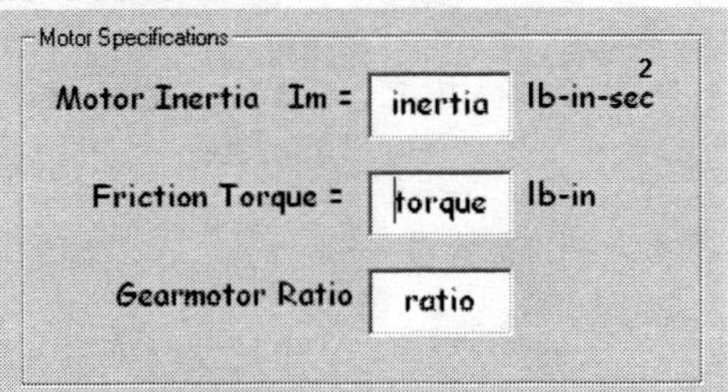

Figure 15. Motor Specifications and Friction.

From this input the computer can calculate the inertia of the system using equation 21.

$$I_{TOTAL} = (QtyofRollers) \cdot I_{ROLLER} + I_{LOAD} \qquad (21)$$

The total inertia of the system is equal to the inertia of the quantity of rollers and the inertia of the load. This reflected inertia[2] at the motor is equal to the system inertia divided by the square of the speed reduction shown in equation 22.

$$I_{REFLECTED} = \frac{I_{TOTAL}}{N^2} \qquad (22)$$

The torque required to accelerate the load and to overcome the bearing friction is shown in equation 23.

$$T_{acc} = I_{REFLECTED} \cdot \alpha_{MOTOR} + T_{FRICTION} \qquad (23)$$

The horsepower required to accelerate or decelerate the load from its initial speed to its required speed is given by equation 24.

$$HP_{SYSTEM} = \frac{T_{acc} \cdot n_{MOTOR}}{63000} \qquad (24)$$

The inertia, torque, and horsepower for a conveyor drive system are calculated and shown in figure 17.

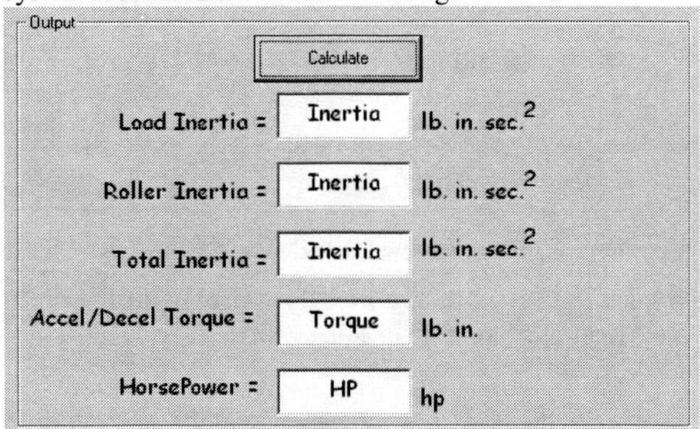

Figure 16. Conveyor Drive Power Requirements.

Leadscrew Drive Analysis Program

The leadscrew drive uses a motor and screw to move a load. The motor is coupled to the screw and rotates the screw. The screw has threads and a mating nut that rides on the threads. The load is attached to the nut. As the screw rotates the nut moves back and forth and moves the load. The power required to move the load is equal to the friction torque and inertia torque of the system. The user can start the leadscrew drive analysis program by selecting the command button shown in figure 18.

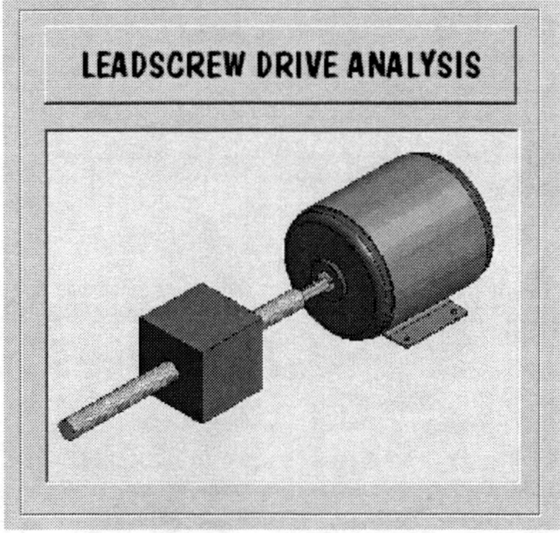

Figure 17. LeadScrew Drive Analysis.

The user inputs the weight of the load be moved by the leadscrew shown in figure 19.

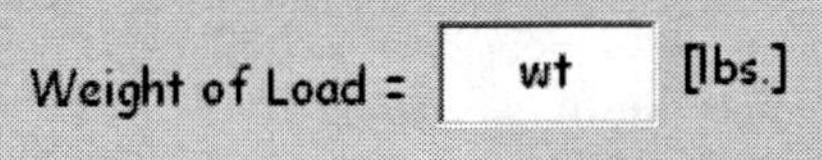

Figure 18. Input Weight of Load.

The user enters the leadscrew coefficient shown in figure 20. The computer program prompts the user to enter this value.

LEADSCREW COEFFICIENTS OF FRICTION

MATERIALS	COEFFICIENT OF FRICTION
Steel on Steel (dry)	.58
Steel on Steel (lubricated)	0.15
Teflon on Steel	0.04
Ball Bushing	.003

Figure 19. LeadScrew Coefficient of Friction.

The friction force is calculated from the weight of the load and the coefficient of friction using equation 25.

$$F_{Friction} = \mu \cdot W_{Load} \quad (25)$$

The user enters the leadscrew efficiency from the table shown in figure 21.

LEADSCREW EFFICIENCIES

TYPE	EFFICIENCY
Ball Nut	0.90
Acme (plastic nut)	0.65
Acme (metal nut)	0.40

Figure 20. Leadscrew Efficiency.

Using the friction force, the leadscrew efficiency, and the pitch of the screw the friction torque can be calculated using equation 26.

$$T_{Friction} = \frac{F_{Friction}}{2 \cdot \pi \cdot p \cdot e} \quad (26)$$

The inertia of the load is calculated using equation 27.

$$I_{LOAD} = \frac{W_{LOAD}}{(2 \cdot \pi \cdot p)^2 \cdot g} \quad (27)$$

The inertia of the leadscrew is based on inertia for a cylinder shown in equation 28.

$$I_{LEADSCREW} = \frac{\pi \cdot L \cdot \rho \cdot R^4}{2 \cdot g} \quad (28)$$

The motor inertia is entered as shown in figure 22.

Motor Rotor Inertia = inertia [lbs. -in. - sec.]

Figure 21. Motor Inertia.

The total inertia can be calculated by summing the load inertia, leadscrew inertia, and the motor inertia reference equation 29.

$$I_{REFLECTED} = \frac{I_{LOAD} + I_{LEADSCREW}}{N^2} \quad (29)$$

The total inertia is used to find the torque required to accelerate the load to the required speed. The acceleration torque is calculated in equation 30.

$$T_{ACCEL} = I_{REFLECTED} \cdot \alpha_{MOTOR} \quad (30)$$

The torque requirement[3] to drive a load using a lead screw is shown in equiation 31.

$$T_{TOTAL} = T_{ACCELERATION} + T_{FRICTION} \quad (31)$$

The horsepower can be calculated using the motor speed and the total torque on the system shown in equation 32.

$$HP = \frac{T_{TOTAL} \cdot n_{MOTOR}}{63000} \quad (32)$$

The leadscrew and load are shown below in figure 23.

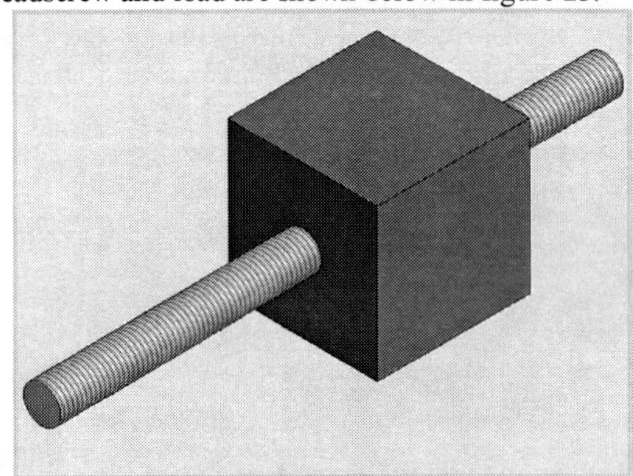

Figure 22. Leadscrew and Load.

The inertia, total torque, and the horsepower requirements for the leadscrew design are calculated in the program and shown in figure 24.

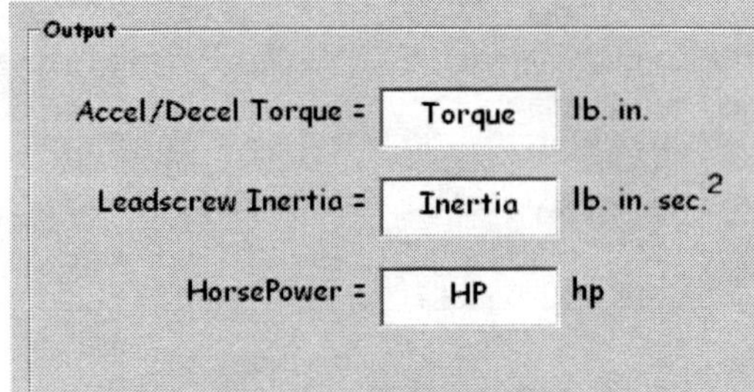

Figure 23. LeadScrew Output.

CONCLUSION

The Computer Aided Design of Power Requirements program will be an interactive program used by students. The program calculates the power requirements of three different drive systems. The program will be used in MET-214 Machine Elements course. This course teaches students to design and analyze machine components. Students learn about torques, forces, and inertias in dynamics. The systems they analyze are usually theoretical and not practical. In Machine Elements, the examples problems have horsepower as a given parameter. Students usually do not have a good understanding of where the power requirements come from. This is an important topic that students need to understand. The program is not a replacement to the traditional teaching and problem solving students do. The program is an additional learning tool. The students should have a good background solving power requirements by hand calculations. Many design problems are iterative in nature. This means the design may change to reduce torques or speeds to influence the design. The advantage of the program is the students can change parameters easily and see the affects on the outputs. It will give the students an opportunity to size many different applications quickly and easily. It also will reinforce the design principles and analytical techniques that are used in calculating the power requirements. The program can also be used in an industrial environment where machine designers are required to calculate the power requirements of different applications.

This program goes one step further than a spreadsheet program. It allows the user to look at pictures, graphs, tables, and charts to view information or to select a value needed in the calculations. Also the program prompts the students to enter values when required. This program is part of a series of Machine Elements programs written in Visual Basic. The other programs are Spur Gear Design, Designing Belt Drive Components, and Design of Ball Bearings.

I am currently in the process of developing a workbook to accompany the programs. This workbook would assist in the use of the programs plus include problems to solve using the programs.

The program has been used in the classroom on a limited bases. So the feedback and assessment of the program on student learning is limited. Students have given positive feedback on the program, but until it is used as an integral part of the course it will not be possible to gage on the results of using the software program in student learning. I have presented the other software programs and have also had a positive feedback from professors teaching in this area. I plan on having the software and a workbook supporting the use of the software available 2004.

Bibliography

1. Jensen/Chenoweth, "Applied Engineering Mechanics", Glenco/McGraw Hill, Ohio, 1983

2. J. Mazurkiewicz, "The Basics of Motion Control", Penton Publishing, Inc., Cleveland, 1996

3. "Engineering Reference", Parker Hannifin, Corporation

Edward M. Vavrek

Edward M. Vavrek is an Assistant Professor of Mechanical Engineering Technology at Purdue University North Central. He has a B.S. in Mechanical Engineering from Purdue, a Masters degree in Mechanical and Aeronautical Engineering from Illinois Institute of Technology, and a Masters in Business Administration from Indiana University Northwest. He has worked as a design engineer for 14 years in the printing industry and steel industry.

Proceedings of IMECE'03
2003 ASME International Mechanical Engineering Congress
Washington, D.C., November 15–21, 2003

IMECE2003-42204

A CASE STUDY IN LABORATORY-BASED ONLINE COURSES - TEACHING CNC PROGRAMMING

James B. Higley, P.E. Purdue University Calumet **David A. McLees, Purdue University Calumet**

Mohammad A. Zahraee, PhD Purdue University Calumet

ABSTRACT

Asynchronous web based instruction, more commonly known as online education or distance learning, has been available for some time. This technology has brought education within the reach of many who would otherwise be unable to attend live classes. Time schedules and distance no longer have a bearing on attending a course. Even group projects are manageable with email and discussion groups. Many courses convert quite well to the online format, and studies have shown that students can learn as much from online courses as from live courses. In many cases, multiple course certificate programs and even some complete degree programs are offered online. As inviting as online courses may be, they do have their limitations, especially classes with a laboratory component. A number of institutions have offered laboratory-based classes in an online format with varying degrees of complexity and success. In some cases, students travel to the institution a limited number of times for extensive laboratory experiences while other institutions use what might best be described as virtual reality based systems. This paper discusses Purdue University Calumet's first laboratory-based online course, MFET 275, Computer Numerical Control Programming Applications. A combination of technologies makes this course successful and effective. Development procedures for this course along with technology used, online pedagogy issues, and course assessment are covered in this paper. Suggestions for future course development complete the discussion.

BACKGROUND

Purdue University Calumet's (PUC's) student base comes from Northwest Indiana and Northeast Illinois including the greater Chicago area. This is a heavily industrialized area with major steel mills in Indiana, several local automotive plants, and a wide variety of Chicago industries including such diverse products as gum, lamps, electronics, plastics, and potato chips. In addition, many smaller companies build and service equipment for the manufacturing companies. Machining takes place in most of these businesses, so a thorough course in computer numerical control (CNC) machining was instituted in the early 1980's to serve those students whose interests lie in automated manufacturing. This is an elective course for students in Mechanical Engineering Technology (MET) and Industrial Engineering Technology (IET) programs at PUC, although engineering students and adult learners also take this course to enhance their careers. The course description follows:

> MFET275, CNC Applications: A study of the principles, techniques, and applications of computer numerically controlled machine tools. G and M code programming of industrial machines, tooling systems, and an introduction to Computer Aided Manufacturing (CAM) systems will be covered.

Many colleges and universities offer similar courses in MET and Manufacturing Engineering Technology programs, e.g. Rochester Institute of Technology [1] and Texas A&M [2]. There are notable exceptions. Purdue University's main campus in West Lafayette, IN teaches some CNC programming in their MET 242, Manufacturing Processes II course for their MET students. Recent laboratory projects for that course include writing a program to machines initials in a CD box lid or writing programs for a two-axis laser cutter. Main campus Purdue Computer Integrated Manufacturing Technology (CIMT) students also learn CNC programming as part of CIMT 244, Automated Manufacturing II. However, no dedicated CNC programming course is offered [3]. These differences simply reflect the needs of each program's constituents with the manufacturing programs having a much higher instance of dedicated CNC courses.

COURSE DEVELOPMENT ISSUES

When taught live, MFET275 meets two hours per week for lecture and two hours per week for lab for a regular 15 week semester. The text for the course is *Computer Numerical*

Control for Machining by Mike Lynch [4]. That along with the *Machinery's Handbook* [5] which the students already have from previous classes, a set of course notes that the students can purchase or print online, and a few floppy disks completes the necessary materials for the course. The students write their CNC programs in any text editor, and then run the programs on either a Haas VF-1 vertical machining center or a Haas SL-20 turning center. See Fig. 1 for a view of PUC's CNC equipment in the Manufacturing Laboratory. The students use Edgecam from Pathtrace Systems to learn computer aided manufacturing (CAM) concepts.

Figure 1 – Haas Turning and Machining Centers

Haas offers as an option a 3.5" floppy disk for their machine controllers, and this option was purchased for both machines. For the small programs used in the class, this provides the fastest and most convenient method of transferring programs. Serial communication may also be used, if desired.

When the decision was made to try MFET275 online, three questions had to be answered:

1. How will the material be presented?
2. Will the students succeed?
3. How will the lab portion of the course be presented?

The answers to the first two questions came easily. Much research has shown online courses to be an effective way for students to learn in a wide variety of disciplines and in different modes, i.e. synchronous or asynchronous. Evans [6] performed an extensive study showing that students in an asynchronous online course learn just as effectively as students in a live course in a graduate level engineering management course. Bhavnani [7] describes a live, internet based graduate course that was offered with good success using live lectures with all participants using video links. The success of the previous two authors, especially since student learning was documented, encouraged this author to think about MFET275 as an asynchronous online course. Gillet's [8] success with a prototype Internet lab-based course provided the needed inspiration to answer question 3 - How will the lab portion of the course be presented? Gillet presents the successful results of a prototype environment where users access remote laboratory experiments via the Internet for five different experiments in the controls area. Those authors prove that laboratory courses can be offered via distance learning format. Although none of those authors addressed the exact situation at PUC, their experiences convinced the author that an online CNC course was feasible. A multimedia combination of audio, static visual, animations, and video was chosen to deliver the course, and the combination provided an effective means of teaching this course. The following paragraphs describe the conversion of MFET275 to distance learning format along with the technologies used.

When converting this course over to the online format, the Lynch textbook and the *Machinery's Handbook* were retained, and the notes were replaced with multi-media internet technologies. The students e-mail programs to the instructor, and the instructor or laboratory technician then transfers the programs to the machine to manufacture parts.

The main delivery vehicle for the online course is Blackboard. This system provides a convenient method for presenting course information and recording student work. While student grades, announcements, and assessment tools are available only to those students and instructors registered in the Blackboard system, it has the flexibility to either house course material entirely or point to material on other servers. By design, the course material is located on the School of Technology server [9]; hence, it is available to all Internet users at all times. This was done to make the material available to other interested parties in hopes of growing one course rather than duplicating efforts.

The course consists of 20 lessons, 10 programs, seven reading assignments, two speed and feed calculation sheets, and a project. The lessons are shown in Table 1.

Table 1 – Online Lessons

Lesson	
1. History and Terminology	11. Rectangular Cycles for Turning centers
2. Introduction to Turning Centers	12. Threading on Turning Centers
3. Speed and Feed Calculations for Turning	13. Programming Arcs
4. Programming Turning Centers	14. Machining Center Example 2 - Arcs
5. Introduction to Machining Centers	15. Hole Cycles
6. Tooling for Machining Centers	16. Hole Cycles with Retract Planes
7. Speed and Feed Calculations for Milling	17. Tool Nose Radius Compensation for Turning Centers
8. Programming Machining Centers	18. Tool Diameter Compensation for Machining Centers
9. Machining Center Example 1	19 Parting on Turning Centers
10. Constant Cutting Speed (CSS) for Turning Centers	20. Roughing and Finishing Cycles on Turning Centers

Each lesson begins with a one to three minute audio clip where the instructor briefly explains what will be covered in the present lesson and why it is important. These audio clips were created using an inexpensive microphone and Soundforge software to record and edit the audio clip. The actual lesson

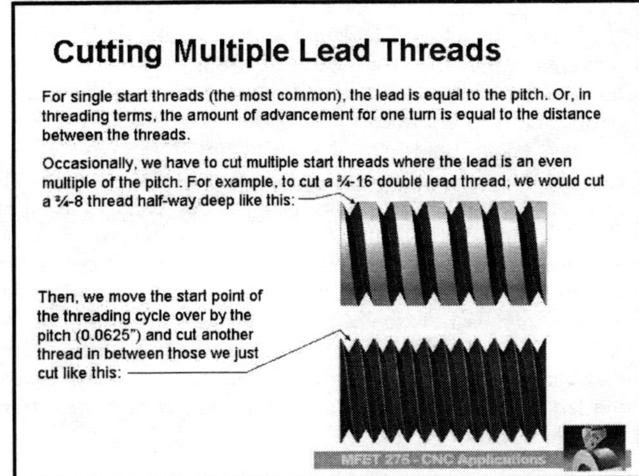

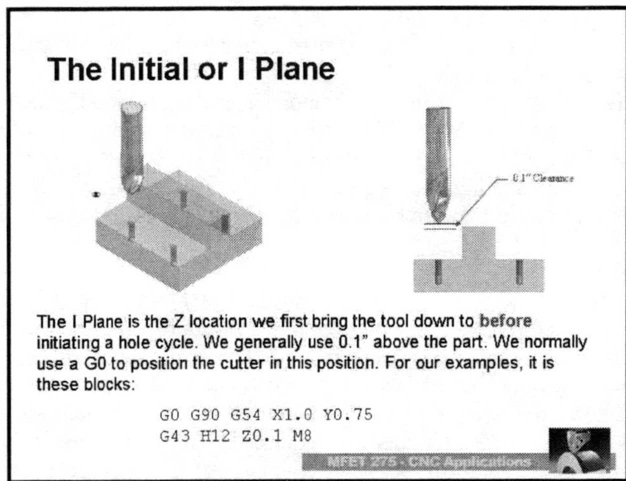

Figure 2 - Sample Turning (Left) and Milling (Right) Slides

material was created in Microsoft Powerpoint. Two sample lesson slides are shown in Fig. 2.

The three-dimensional graphics were created using Autodesk's Inventor product. Images were oriented and then saved as bitmaps and then imported into Powerpoint. The graphics shown in Fig. 2 were created using this method. Some Inventor images were exported to AutoCAD, modified, and then brought into Powerpoint; an example is given in Fig. 3.

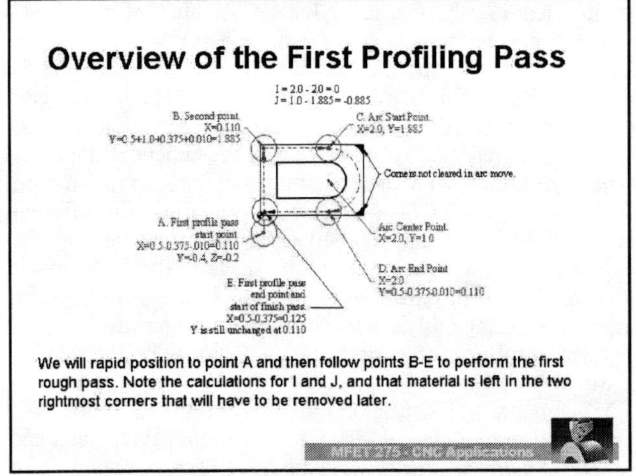

Figure 3 - Inventor Graphics Modified with Autocad

All programming examples also include an animation of the machine's movement. All but one animation were created using Inventor's presentation mode. By assembling the cutter to the part and then tweaking the cutter, basic moves are fairly easy to represent. This was probably not what Inventor's developers intended, but it works reasonably well. The biggest shortcoming is that rapid and feed move speeds cannot be distinguished in the animation. The roughing and finishing cycle animation could not be done in Inventor, so the parts were exported to 3D Studio, and the animation was done with that product. A sample view of an Inventor animation is shown in Fig. 4.

Since animations require considerable time to load, each animation was converted to streaming video and then all references to the animations were linked to the streaming video server. This makes the animations load and run acceptably even from a 56K modem.

Since students did not personally attend the laboratory, each program was videotaped with a digital video camera, edited in digital format using Adobe Premier 6.0, converted to streaming video, and then linked to the course web page. Some video image quality was lost, but the quality was certainly acceptable, and the videos run well even on a 56K modem connection.

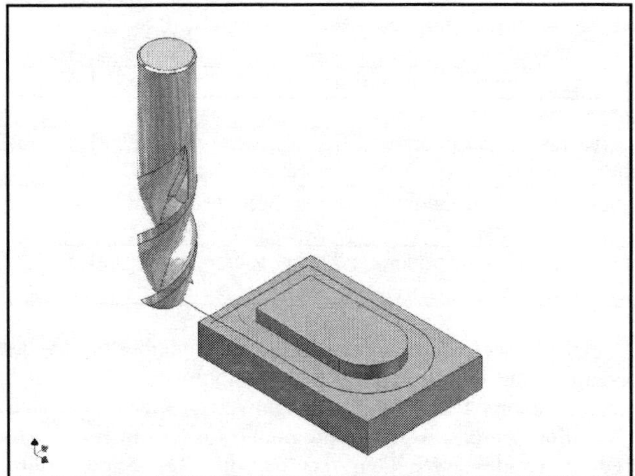

Figure 4 - Sample View of an Inventor Animation

ONLINE PEDAGOGY ISSUES

The Organizational Leadership & Supervision (OLS) Program at PUC has been a campus leader in online course development. Their first course was offered in 1998, and they currently have 14 courses available including a five course Safety Certificate. Hence, the METS Department has a considerable experience base concerning online courses and potential problems. Much of the following discussion is based on suggestions from the OLS faculty.

The OLS faculty warned that material preparation for an online course usually runs between 300 and 500 person-hours. Since a small grant from the School City of Hammond was used to fund the MFET275 online course development, a detailed time log was kept. Approximately 210 hours were required for creating the course materials and approximately 145 hours were required to convert the material into fast loading online format. Hence, this course fit the OLS experience.

Previous experience proves students seldom read the course syllabus nor follow the course procedures. To insure initial success, the department mails a letter to each student during the week before class with instructions for accessing Blackboard and setting an active email address. During the first week of class, the students are required to take and pass a syllabus quiz. The quiz does not count towards the student's final grade, but the student must pass the quiz with a 100% score. This insures that the students are familiar with course policies. The questions and response types are shown in Table 2.

Table 2 – Questions from the Syllabus Quiz

Question	Response Type
Professor Higley's office is CLO346	True/False
This course covers computer numerical control of machine tools.	True/False
You will learn how to use G and M codes to program CNC machine tools in this course.	True/False
To earn a B grade in this course, my final score must be between:	Multiple Choice
Match the following course assignments with how much they contribute to your semester grade.	Matching
The final exam is comprehensive.	True/False
Late work is always accepted.	True/False
I have read and understand the laboratory safety rules.	True/False
I have read and understand the Student Responsibilities.	True/False
I have read and understand the course schedule.	True/False

After successfully completing the syllabus quiz, the students could then start on weekly assignments. They simply emailed completed reading assignments, speed and feed calculations, and all CNC programs to the instructor for grading. Grades were then recorded in Blackboard for the students to view. Rather than write comments on each student's work, correct assignments and programs were posted in Blackboard for the students to view. Some students called or emailed questions, and these were handled individually.

Given the format of the course, the midterm and final exam have to be open book, open note, but Blackboard has a timer, and time was limited to two hours for each exam. Approximately 10-15 multiple-choice questions were included along with two programs for each exam. Images are easily imported into Blackboard, so the instructor provided a multiview drawing and a tooling list, and the students typed their programs into the test. Blackboard grades the multiple choice questions automatically, but the instructor must go in and assign points for each program. Once the points are assigned, Blackboard calculates the final grade. Nothing prevents more than one student from sitting together on multiple computers and taking the test together. Given the non-traditional nature of most of the students in this first class, this risk was taken. Based on the time stamps from each completed test and the variations in grades, the instructor was satisfied that no cheating took place. This is a serious concern for larger classes. The OLS faculty are using Testpilot which creates individual tests for each student by pulling from a large question bank in each of the topic areas. Hence, even if students are sitting right next to each other, chances are slight that they will have exactly the same test. This works well with questions that can be automatically graded, but poses a heavy grading burden when an instructor must grade a dozen or more different types of CNC programs.

COURSE ASSESSMENT

Two types of course assessment take place: the instructor's assessment of the students and the student's assessment of the course. The first type was covered in the previous section as it is integral to pedagogy issues. This section discusses student assessment of the course (now referred to as course assessment) and then presents a method of presenting both the instructor and student course assessment in a form useful for continuous improvement efforts.

In the fall 2002 semester, PUC began a trial of an online course assessment tool using the survey function in Blackboard. With this feature, the instructor knows if a student performed the assessment, but all student answers are grouped together so individual student responses remain anonymous. This tool is broken down into four parts: Student Self-assessment, General Course Impact (ABET concerns), Course Management, and Course Objectives. Shown in Table 3 in the appendix, this tool has been designed with the first three sections common, and with the course objectives portion easily modified for different courses. The appropriate Lichert scale for each question has been removed in the interest of space. Although much modified, it is based on the work of Land and Hager [10]. The course assessment tool is part of a larger project to perform integrated, on-line assessment of all courses in the METS Department.

To compare the instructor's assessment of the course with the students', an Excel spreadsheet consisting of three parts has been developed to track the data. See Tables 4-6 in the appendix. Much research on course assessment tools of this type is available [11], and this is among the simpler types. The author intentionally created a simple form because these forms must be generated for 23 courses at the same time, and the faculty felt a short, simple form would be the best place to start. After results have been analyzed from a year or two's data, the forms will be revisited.

Part 1, shown in Table 4, lists the scores from specific assignments that the instructor uses to measure each course objective. The students' evaluation of how well they felt the course met each objective is listed as well. Part 2, shown in Table 5, lists the students' perception of how well the course met the ABET a-k criterion. The last part, shown in Table 6, provides a place for the instructor to record course changes and improvements. The METS Department plans on having similar forms for each course and then linking the data to the web page for each course. This will provide a convenient method for storing course data and making it easily available to instructors and ABET teams.

CONCLUSION

With 13 students enrolled in MFET275 in its online format, statistically meaningful data is not available. However, midterm and final exam scores were within normal ranges for this course, programming lab scores were slightly lower than for live classes, the project to research CAM systems was quite successful, and student evaluation of the course was quite good. The author feels this online course provided learning almost as good as a live course. With some improvements, equal learning should be possible.

ACKNOWLEDGMENTS

The author would like to acknowledge the School City of Hammond for providing a grant to help fund the creation of this course.

REFERENCES

[1] Rochester Institute of Technology Web Page, http://www.rit.edu/~719www/

[2] Texas A&M Web Page, http://etidweb.tamu.edu/manufacturing/courses.html

[3] Purdue University MET Department Web Page, http://www.tech.purdue.edu/met/

[4] Lynch, Mike, 1992, *Computer Numerical Control Machining*, McGraw-Hill, New York.

[5] Oberg, E., Jones, F. D., Horton, H. L., and Ryffell, H. H., 2000, *Machinery's Handbook*, 26th Edition, Industrial Press, New York.

[6] Evans, Rosemarie M., Murray, S. L., Daily, M., and Hall, R., 2002, "Effectiveness of an Internet-Based Graduate Engineering Management Course," *Journal of Engineering Education*, **89**, no. 1, pp. 63-71.

[7] Bhavnani, S. H., Bar-Cohen, A., and Joshi, Y. K., 2000, "The Classroom of the Future: An Internet-Delivered National Course on Thermal Management of Electronics," *Journal of Engineering Education*, **89**, no. 4, pp. 423-427.

[8] Gillet, D., Latchman, H. A., Salzmann, C., and Crisalle, O. D., 2002, "Hands-On Laboratory Experiments in Flexible and Distance Learning," *Journal of Engineering Education*, **90**, no. 2, pp. 187-191.

[9] MFET275 course material, http://technology.calumet.purdue.edu/met/mfet/275/

[10] Land, R. and Hager, W., 2002, "Pilot Survey: Graduate Satisfaction with ET Education at Penn State," *Proceedings of the 2002 ASEE Annual Conference*, Montreal, June 16-19.

[11] Henderson, B., Jeruzal, C., Pourmovahed, A., 2002, "Assessment of Student Cognitive Development in the Energy Systems Laboratory," *Proceedings of the 2002 ASEE Annual Conference*, Montreal, June 16-19.

APPENDIX

Table 3 - Student Assessment Tool

Student Assessment Tool
Specific Student Responsibility Questions: 1. I attended scheduled classes and labs. 2. I arrived on time for scheduled classes and labs. 3. I read the course material/text when it was assigned. 4. I was well prepared for class. 5. I participated in classroom discussions and activities. 6. I used the supplemental materials or website (Bb) my instructor provided. 7. My ability to apply knowledge from pre-requisite courses for this course can be rated as,
General Course Impact Questions: 8. As a result of this course, my mastery of the knowledge, techniques, skills, and modern tools of the Mechanical Engineering Technology discipline can be described as, 9. As a result of this course, my ability to apply current knowledge and adapt to emerging applications of mathematics, science, engineering, and technology can be rated as, 10. As a result of this course my ability to conduct, analyze, and interpret experiments and apply results to improve processes can be rated as, 11. As a result of this course, my ability to apply creativity in the design of systems, components, or processes appropriate to program objectives can be rated as, 12. As a result of this course, my ability to function effectively on teams can be rated as, 13. As a result of this course, my ability to identify, analyze, and solve technical problems can be rated as, 14. As a result of this course, my ability to communicate effectively can be rated as, 15. As a result of this course, my recognition of the need for, and an ability to engage in lifelong learning can be rated as, 16. As a result of this course, my ability to understand professional, ethical, and social responsibilities can be rated as, 17. As a result of this course, my respect for diversity and knowledge of contemporary professional, societal, and global issues can be rated as, 18. As a result of this course, my commitment to quality, timeliness and continuous improvement can be rated as,

Specific Course Management Questions:

19. My instructor passed out a syllabus or made one available in the Internet early in the course.
20. I was able to understand the syllabus and grading procedures.
21. The instructor followed the syllabus.
22. Given the ease or difficulty of the material presented in this course, the exams represented the topics covered fairly.
23. The course assignments were related to the material being covered.
24. The laboratory assignments in this course help reinforce the topics being covered and make them easier to learn. (Only for classes with labs.)
25. My instructor returned graded material such as homework and tests in a timely manner.
26. My instructor was on time and prepared for class.

Specific Course Objective Questions:

27. A specific objective of this course is to explain the terminology used to describe CNC machine tools. How well did this course meet this objective?
28. A specific objective of this course is to explain the basic types of CNC machine tools and the manufacturing operations for which they are best suited. How well did this course meet this objective?
29. A specific objective of this course is to describe the major factors in the development of CNC machines. How well did this course meet this objective?
30. A specific objective of this course is to explain the tooling requirements of CNC systems. How well did this course meet this objective?
31. A specific objective of this course is to prepare G and M code programs and documentation for the manufacturing steps required to produce machined parts on CNC turning and machining centers. How well did this course meet this objective?
32. A specific objective of this course is to describe the importance of product design and tool design for the successful production of a product. How well did this course meet this objective?

Table 4 – Course Objective Assessment Tool

MFET275 CNC Applications Course Assessment Tool – Instructor and Student Course Objective Assessment												
Semester: Fall 2002				**Instructor: Higley**								
		Course Embedded Assessment of Student Performance						**Student Evaluation (%)**				
Course Objective	**Supported Related Outcome**	**Assessment Tool 1**	**Score (%)**	**Assessment Tool 2**	**Score (%)**	**Assessment Tool 3**	**Score (%)**	**E**	**G**	**A**	**P**	**NA**
1. Explain the terminology used to describe CNC machine tools.	1.1 Technical Proficiency	Midterm Questions 1, 6, 7	70					56	11	33	0	0
2. Explain the basic types of CNC machine tools and the manufacturing operations for which they are best suited.	1.1 Technical Proficiency	Final Questions 1, 2	100					33	44	22	0	0
3. Describe the major factors in the development of CNC machines.	3.2 Human Differences	Midterm Questions 4, 5	89	Final Question 3	62			33	33	33	0	0
4. Explain the tooling requirements of CNC systems.	1.1 Technical Proficiency	Midterm Questions 10, 11, 15	74	Final Question 4	75			22	44	33	0	0
5. Prepare G and M code programs and documentation for the manufacturing steps required to produce machined parts on CNC turning and machining centers.	1.1 Technical Proficiency 1.3 Computer Applications 1.4 Open-ended Problems	Lab Assignments	80	Midterm	76	Final	74	44	11	44	0	0
6. Describe the importance of product design and tool design for the successful production of a product.								11	33	56	0	0
Instructor Comments for needed changes: Objective 6 is not really covered in class, and it has low student ratings. I'll change this objective to include CAM systems for the next class.								Number of responses: 9				

Table 5 – ABET Criterion Assessment

MFET275 CNC Applications Course Assessment Tool ABET Criterion Assessment Proposed ABET Criterion Satisfied: a,b,d,f,j,k						
Instructor: Higley	**Semester: Fall 02**					
	Student Evaluation (%)					
Criteria	**E**	**G**	**A**	**P**	**NA**	**Composite**
a. As a result of this course, my mastery of the knowledge, techniques, skills, and modern tools of the Mechanical Engineering Technology discipline can be described as,	0	44	33	0	22	2.97
b. As a result of this course, my ability to apply current knowledge and adapt to emerging applications of mathematics, science, engineering, and technology can be rated as,	0	44	44	0	11	3.19
c. As a result of this course my ability to conduct, analyze, and interpret experiments and apply results to improve processes can be rated as,	22	33	33	0	11	3.52
d. As a result of this course, my ability to apply creativity in the design of systems, components, or processes appropriate to program objectives can be rated as,	11	44	33	0	11	3.41
e. As a result of this course, my ability to function effectively on teams can be rated as,	11	22	44	0	22	2.97
f. As a result of this course, my ability to identify, analyze, and solve technical problems can be rated as,	11	56	33	0	0	3.78
g. As a result of this course, my ability to communicate effectively can be rated as,	22	33	33	0	11	3.52
h. As a result of this course, my recognition of the need for, and an ability to engage in lifelong learning can be rated as,	33	22	44	0	0	3.85
i. As a result of this course, my ability to understand professional, ethical, and social responsibilities can be rated as,	11	22	44	0	22	2.97
j. As a result of this course, my respect for diversity and knowledge of contemporary professional, societal, and global issues can be rated as,	0	22	44	11	22	2.64
k. As a result of this course, my commitment to quality, timeliness and continuous improvement can be rated as,	33	33	33	0	0	3.96
Instructor Comments: Some student evaluations make little sense such as e since no team assignments were included in this class. However, all students agreed that this class supported criteria k, commitment to quality, timeliness, and continuous improvement. Perhaps the distance learning format helps students in those areas.						

Table 6 – Course Update Form

MFET275 Instructor Update Information			
Date Submitted: 12-16-02		**Date to be Reviewed: 01-04-04**	
Responsible faculty for the review: Higley			
Type of Update			
☐ New Edition of the Text	☐ New Text Adopted	☐ New Software	☑ Teaching Method
☐ New Laboratory Equipment	☑ Lab Material Update	☑ Teaching Initiative	☐ Other
Description of Condition Prior to / After Update: This course was taught online for the first time during the fall 2002 semester with good results (see previous pages of assessment data). I will have to require the students to visit the lab and watch parts being manufactured rather than just relying on the video.			
Assessment Method Used to Evaluate Short or Long Term Results: The course evaluations were good for an online class, and I will change the objectives as mentioned on the first page and add live lab time as mentioned above.			

Proceedings of IMECE'03
2003 ASME International Mechanical Engineering Congress
Washington, D.C., November 15–21, 2003

IMECE2003-42382

Thermal-Fluid Sciences Laboratory Course

Ella Fridman, University of Toledo

Maruthi Devarakonda, University of Toledo

ABSTRACT

At the Department of Engineering Technology at the University of Toledo we are working on introducing laboratory course in thermal-fluid sciences with the elements of controls. The objective is to create a course where the students will learn to see connections between the three branches of the thermal-fluid sciences and will learn to appreciate the integrated nature of modern systems.

The proposed laboratory course will be structured around four "equipment centers" represented important areas of applications in the thermal-fluid sciences: 1) Armfield Modules (turbomachinery), 2) Steam Motor and Energy Conversion Set, 3) Modular Air Flow Laboratory, and 4) Experimental Heat Pump and Air Cooler. The students will take the course after they finish courses in Thermodynamics and Fluid Mechanics. The course will give our students an opportunity to observe how the concepts and principles learned in the classroom can be applied in a practical situation. To extend the instructional capability of the equipment and to give the students knowledge of modern instrumentation and control systems we use LabVIEW for computerization and automation of this laboratory.

The experiments that have already been developed for the course as well as experiments that have been planned for the future are presented in the paper.

INTRODUCTION

The rapid rate of change in the fields of technology poses special problems for academic institutions, specifically for the engineering disciplines [1]. There is a continuous need to update and augment the content of laboratory lecture courses to track the change. The laboratories bring the course theory alive so that the students can understand how natural phenomena affect the real world. That is why in the case of engineering disciplines, it is an essential requirement to educate the students with meaningful and relevant course work with major emphasis on experimentation [2]. One of the ways to expose the students to real time engineering problems is to use the computer-based techniques with suitable front-end tools [3]. A computer equipped with instrumentation software and data acquisition hardware supported with suitable interface circuits would enable a student to understand and appreciate the real time experimental modules. Though a number of computer based customized engineering courses and test systems have been developed to replace the conventional mode of instruction in engineering laboratories, they pose a number of limitations such as inflexible and unfriendly programming structure and long term incompatibility issues. But due to the recent advancements in software and computer technology, it is now feasible to implement more efficient, highly interactive and highly user-friendly systems without using expensive custom written software and tools. The concomitant development of computer based data acquisition systems, equation solving and spreadsheet software has provided engineering educators with a plethora of opportunities to positively and creatively impact course content, pedagogy and student interest.

The Department of Engineering Technology at the University of Toledo is taking a leap in educating the undergraduate mechanical engineering technology students by introducing a new inter-disciplinary thermal-fluids laboratory course. This course integrates team oriented, hands on learning experiences throughout the engineering curriculum. The curriculum of this course at the University not only emphasizes the design and open-ended problem solving elements necessary to a competitive engineering education, but seeks to augment the communications and leadership skills which the department has embraced as desired outcomes. The authors would like to envisage an integrated approach [4] in the development of the course bringing in the thermal sciences, fluid mechanics, and instrumentation and control system aspects of Mechanical Engineering, thus enabling the student to encounter real world situations in the laboratory. The course is being designed to provide the students with a combination of lecture and laboratory experiences combining the concepts of Thermal Sciences with the topics of data acquisition, statistical analysis, Fourier analysis calibration, and operation of transducers. Efforts are being put in to develop a Learning factory [4] that proposes a novel approach to the shared laboratory experience that uses interdisciplinary courses and multidisciplinary teams to link the course laboratories in a logical sequence leading to an integrated learning environment approach that enhances student learning. This approach redefines the way laboratory instruction is designed and enables the instructors to focus on long-term goals. The students can appreciate the novel concept of vertical integration where they need to utilize their previous laboratory work experience as building blocks for the setups in multiple engineering laboratories.

1. Design and Development plan

The thermal-fluid sciences (thermodynamics, fluid mechanics, and heat transfer) are typically considered to be among the most difficult topics taught in an engineering curriculum [5]. In this project it is our intent to improve the students' understanding of this subject matter by implementing an integrated experimental facility for applications in thermo-fluids. In this facility we are planning to utilize the same experimental set-ups and apparati to demonstrate phenomena from the three points of view of thermodynamics, fluid mechanics and heat transfer. In this way, the students will learn to see the nexus between the three branches of the thermal-fluid sciences and appreciate the integrated nature of modern systems.

The thermal fluid sciences laboratory is going to be structured initially around four "equipment centers". These centers represent the areas of application in thermal fluid sciences which is an integrated discipline comprising Thermodynamics, Heat Transfer and Fluid Mechanics. In order to enable the student appreciate the underlying concepts of the real world engineering problems, it has been decided to integrate and update the conventional experiments with custom written instrumentation software thus establishing a virtual laboratory in the future. A virtual laboratory is an interactive environment for creating and conducting simulated experiments. There are several criteria that were contemplated for the selection of instrumentation software to build virtual laboratories such as modularity, multi platform portability, and intuitive graphical user interface and advanced debugging features. But finding a software tool that provides all the functions to achieve these tasks is very difficult and so research has been done into different instrumentation software used in engineering education. Based on a survey [1] conducted for a suitable selection of teaching and learning tools in engineering disciplines, National Instruments LabVIEW has been chosen to extend the instructional capability of the equipment and educate the students in instrumentation and control systems. It is a graphical programming language that allows engineers and scientists to develop customized virtual instrument, which is flexible, modular and economical. This software does data acquisition, analysis and control in addition to the multimedia authoring capabilities with the help of add on tools.

2. Laboratory course Scope and Objectives

The thermal-fluids laboratory course is being designed with the following objectives.

1. To provide the student an opportunity to apply the concepts presented in thermal and fluid sciences.
2. To familiarize the students with measurement techniques using "state-of-art" instruments that are encountered in industry.
3. To give the student hands-on experience in computer data acquisition, analysis and control systems
4. To establish a process of continuous improvement and development of the course by enabling the students to design a new experiment for one of the equipment centers as a part of the course final project, following the concept of vertical integration [4].

3. Modular Air flow laboratory Experiment

To illustrate the integrated nature of the experiments in the thermal fluids laboratory course, an experiment with one of the four equipment centers is presented. The center is a modular air flow laboratory developed by TQ Education and Training Ltd.

3.1 Equipment

The Modular air flow laboratory is a compact, vertical axis, low speed, blower type wind tunnel with a working section of 185 mm x 100 mm. It includes a fan with single phase motor, inlet valve for wind speed control, reference pressure tapings for air speed measurement, pilot traverse and vertical manometer, plain working section window, mounting working section window with protractor. Different geometric models such as airfoil, cylinder and sphere are available at the equipment center to provide a wide range of experiments in heat transfer, fluid mechanics and aerodynamics.

Fig 1. Modular Air Flow Laboratory [Courtesy: TQ Ltd]

3.2 Objectives

- To calculate the coefficient of lift on airfoil by creating a virtual instrument for measuring the pressure distribution on the airfoil.
- Verify the pressure data obtained by computer based instrumentation with the data obtained manually.
- Validate the data obtained using a known scientific approximation.
- Analyze and study the impact of different factors, such as icing, on pressure distribution and lift coefficient in an airfoil.

3.3 Programming a Virtual Instrument

A comprehensive tutorial is drafted and is available for the students to enable them program a virtual instrument. Given below is the diagram window of the virtual instrument developed for pressure data acquisition.

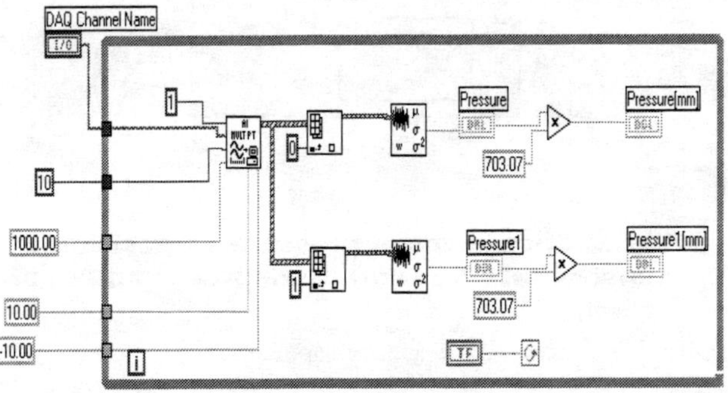

Fig 2 Diagram window of the Virtual Instrument

3.4 Description of the Experiment

The air foil used in the experiment is provided with 12 static pressure tapings to enable the measurement of pressure and lift coefficients. To obtain the pressure on the airfoil, the aluminum rear wall is mounted on the back of the working section with the hole near the contraction. The airfoil is aligned with the support rod through the bush in the Perspex window with a protractor. The pressure tubes are passed through the hole in the working section rear wall connecting the air foil pressure tapings to the pressure transducer PX138-0.3D5V [Courtesy: Omega Technologies].

The pressure transducer is chosen because of the low pressure ranges it can measure with a negligible error. The differential pressure range that can be measured with the transducer is ±0.3 Psia corresponding to a voltage of around 1-6 V at an excitation voltage of 8V. The pin-out configuration of the pressure transducer is shown below.

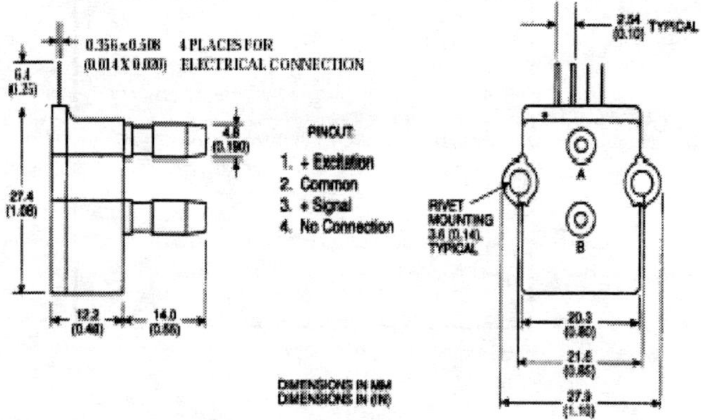

Fig 3 Pin out diagram of the transducer [Courtesy: Omega Corporation]

The Pressure signal at the transducer is measured with respect to a common negative terminal. The output of the transducer is connected to the National Instruments data acquisition card PCI 6024E through the SCB-68 Connector block. The obtained data is displayed on the front panel of the instrument and thus used for the calculation of pressure and lift coefficients.

3.5 Calculation of Lift coefficient

To calculate the lift coefficient on the airfoil, the graph on Fig 4 is considered. The graph presents the pressure coefficient with respect to the position of sensors in terms of percentage from the nose of the airfoil along the chord. The upper curve represents the pressure on the lower side of the airfoil and the lower curve represents the pressure on the upper side. The area between these curves defines the value of the lift coefficient [C_L], because it is the difference in pressure lying on these surfaces that determines the lift of the airfoil. As the graphs are plotted considering the position of sensors as percentage of the airfoil, the area is divided by 100 to give the lift coefficient.

3.6 Validation of data

The validation of data is done by comparing the experimental and theoretical values of lift coefficient at zero angle of attack. The plot of pressure coefficient Cp at different points on the airfoil at zero angle of attack is shown below.

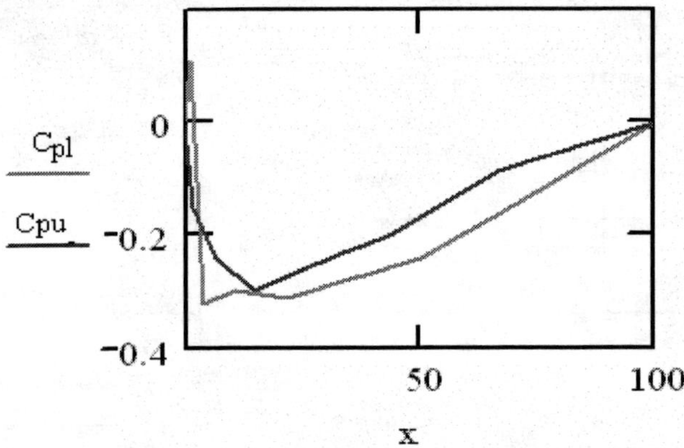

Fig 4 Coefficient of pressure vs position of sensors on the airfoil

On the graph, Cpl – pressure coefficient on the lower surface of the airfoil; Cpu – pressure coefficient on the upper surface of the airfoil; x– distance along the chord from the nose of the airfoil in percentage

The area between the curves in the plot is found to be 2.5 Square units. This gives the coefficient of lift at zero angle of attack C_L = 0.025, which is close to the theoretical value of C_L=0. The relative error in the calculation of C_L can be estimated as

$$\frac{\text{Difference in area between the curves}}{\text{Area under the curve}} = \frac{2.5}{40} = 6.25\,\%.$$

Experimental data and graphs for angles of attack 2.5 and 5 are presented below.

Table 1: Coefficient of pressure at different angles of attack

			0	0	2.5	2.5	5	5
Position of Sensors as % of airfoil			Cp Lower	Cp Upper	Cp Lower	Cp Upper	Cp Lower	Cp Upper
L	1		0.108		0.599		0.688	
U		2		-0.149		-0.839		-1.155
L	4		-0.317		0.071		0.201	
U		7		-0.24		-0.607		-0.74
L	11		-0.291		-0.04		0.041	
U		15		-0.295		-0.548		-0.639
L	22		-0.306		-0.13		-0.06	
U		25		-0.261		-0.42		-0.478
U		44		-0.196		-0.289		-0.322
L	50		-0.24		-0.15		-0.12	
U		66		-0.156		-0.2		-0.218
L	80		-0.1		-0.06		-0.05	

L – Lower side of the airfoil
U – Upper side of the airfoil

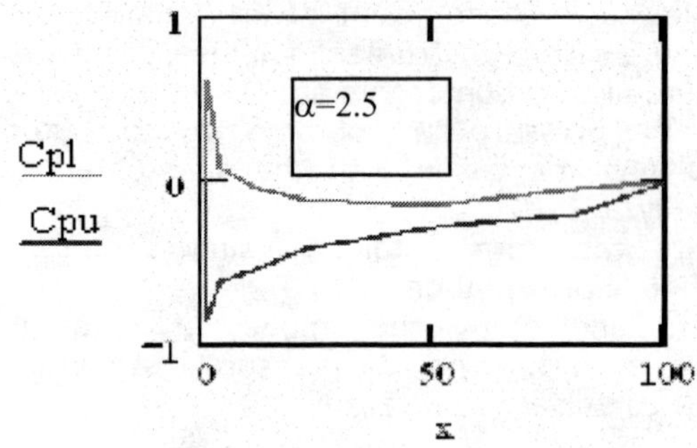

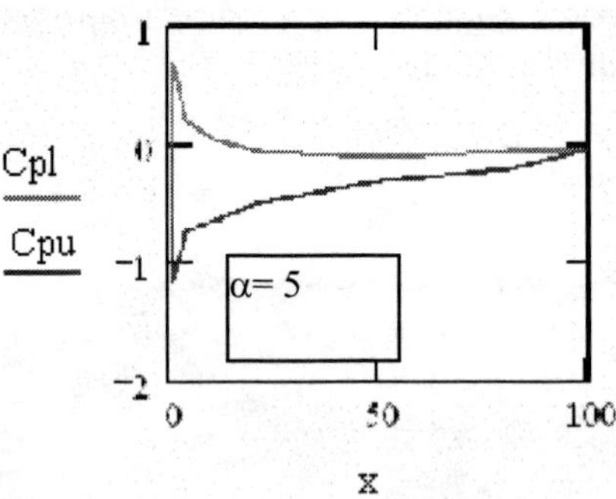

Fig 5. Coefficient of pressure vs position of sensors at 2.5 and 5 degree angles of attack

4. Results and Discussion

The integrated nature of one of the equipment centers in the thermal-fluids course illustrated earlier would enable the students to develop a deeper insight into the aerodynamic concepts simultaneously enhancing their instrumentation skills. The essential electronics behind the data acquisition system and the concepts of aerodynamics would act as a firm framework to develop their own final project using this experiment, based on the approach of vertical integration. Numerical modeling of the fluid flow at different conditions using computational fluid flow packages is one of the future tasks in the considered equipment center. The authors envisage a highly productive and novel outcome by exposing students to common engineering software packages especially in automotive and aerospace industry. The students would be able to meet the demands of the industry in terms of the experience the course offers in various disciplines of the thermal-fluid sciences.

5. Conclusions

The authors would like to present here the learning outcomes of the course. With the completion of this course, the students will be able to

a. Demonstrate knowledge of engineering concepts that are fundamental in the areas of thermal-fluids sciences and data acquisition systems.
b. Identify, analyze and solve technical problems pertaining to the design of data acquisition and control systems in thermal-fluids sciences using the principles of calculus and engineering science with the appropriate usage of computer technology.
c. Plan, conduct and interpret experiments and apply the experimental results to simulate the vertical integration approach.
d. Apply creativity in designing new experiments with the available equipment centers continuously improving the course

REFERENCES

1. Nesimi Ertugrul, *Towards Virtual Laboratories: a survey of LabVIEW based teaching/Learning tools and future trends* – International Journal of Engineering Education, Vol 16, No.3 pp, 171-180, 2000.
2. Bordogna, J., Fromm, E., and Ernst E., *Engineering Education: Innovation through Integration*, Journal of Engineering Education, 82, 1, January 1993
3. National Instruments Developer Zone, *Distance learning remote laboratories using Labview* - National Instruments 2000.
4. Ajay Mahajan, Maurice Walworth, David McDonald and Kevin Schmaltz: *Integrated Systems Engineering Laboratory – An Innovative Approach to vertical Integration using modern instrumentation,* session 2259, American Society of Engineering Education, 1999.
5. Mahajan Ajay and David McDonald, "An Innovative Integrated Learning Laboratory Environment", presented at ASEE Annual Conference, June 1997.
6. www.ni.com
7. Modular Air flow laboratory user manual.

Proceedings of IMECE'03
2003 ASME International Mechanical Engineering Congress
Washington, D.C., November 15–21, 2003

IMECE2003-55527

AN INNOVATIVE FUEL CELL THEORY, TESTING AND MANUFACTURING COURSE

Brad Rogers
Arizona State University – East

Govindasamy Tamizhmani
Arizona State University – East

Rajiswari Sundararajan
Arizona State University – East

Scott Danielson
Arizona State University – East

ABSTRACT

A unique course in the theory, testing and manufacturing of fuel cells has been developed at Arizona State University in the College of Technology and Applied Sciences. The course is designed for the engineering technology graduate students, but is also accessible to advanced undergraduates. The interdisciplinary nature of the course necessitates a team-teaching approach, and faculty with backgrounds in electrochemistry, electrical engineering and mechanical engineering deliver portions of the course. The course includes a theoretical portion, but also contains a comprehensive practical portion in which the students build a membrane electrode assembly and assemble, test and characterize this assembly as a single stage proton exchange membrane fuel cell. Feedback from the students indicates a great deal of excitement over the course, and has resulted in several students deciding to concentrate their graduate work in the fuel cell area. The paper describes the first course offering in the spring of 2003.

INTRODUCTION

Fuel cells are a well known technology for the direct and efficient production of electrical energy from chemical sources. The most efficient fuel cells operate using hydrogen as the fuel, resulting in essentially no pollution from the electrical generation process. Although the operation of fuel cells was first demonstrated in 1839[1], their use in the production of electrical power has not been widespread due to the relative simplicity of systems relying on heat engines. More recently, however, concerns over environmental deterioration and the location and size of remaining fossil fuel reserves, as well as increasing costs, have led to intense interest and investment in the development of hydrogen fuel cell technologies, both in the public and private sectors. For example, in his January 28, 2003 State of the Union address, President Bush recommended the commitment of 1.2 Billion dollars in research funding for the development of hydrogen technologies[2]. Most major automotive manufacturers have fuel cell development programs underway, and have produced prototype vehicles for the consumer market[3]. The stationary fuel cell market, including residential applications, is projected to grow to 40 Billion by the year 2010[4].

The potential for success of fuel cells is resulting in the growth of new industries to support and develop the technology. However, there are a great many unresolved problems that are preventing large scale market penetration of this technology. While these problems are seemingly tractable, their solution will ultimately require the availability of scientists and engineers with an understanding of both the theory and application of fuel cells and related technologies so that they may focus their talents directly to these problems. Higher education bears a responsibility to respond to this need. With this in mind, a fuel cell course has been developed at ASU East, targeted to graduate students and advanced undergraduates, in a cooperative effort between the departments of Electronic Engineering Technology, Mechanical and Manufacturing Engineering Technology, and the Fuel Cell Laboratory.

A particular difficulty in the delivery of an effective fuel cell course is the interdisciplinary nature of the field. The fundamental process by which hydrogen reacts with oxygen to produce water, heat and electricity is an electrochemical process that has been studied and perfected by electrochemists. The production and transport of hydrogen, the mechanical and thermal design of the systems, and the manufacturing of the system components is largely the field of mechanical engineering. The direct power conditioning, inversion to AC power, process controls, and associated equipment such as batteries and motors fall under the field of electrical engineering. The practical knowledge and difficulties of assembly, testing and integration of components requires expertise in the laboratory. Effective fuel cell development and production will require a team of experts from different disciplines, working together toward a common goal. For individual specialists to be productive members of this team, they must understand the interdisciplinary nature of the field, including the problems and constraints that other specialists are faced with. Therefore, the first course in a graduate fuel cell program should embrace the interdisciplinary nature of the field, and provide this perspective, including laboratory experiences. This course can then be effectively followed by more specialized topics suited to individual backgrounds and interests.

COURSE INSTRUCTORS

The course was team-taught by three faculty from ASU East. Dr. Govindasamy Tamizh-Mani is an electrochemist with experience in fuel cell research, including the development of catalyst for fuel cell electrodes. Dr. Rajeswari Sundararajan is an electrical engineer with experience in power conditioning and control. Dr. Brad Rogers is a mechanical engineer with experience in the development of fuel cell accessory systems.

STUDENT CROSS SECTION

21 students registered for the course in the Spring of 2003. Of these, 13 were graduates students in the College of Technology and Applied Sciences at ASU East, and 8 were seniors in this college. Among the Graduate students were 6 from the Electrical Engineering Technology (EET) program, 6 from the Mechanical Engineering Technology (MET) program, and 1 from the Global Technology and development program. The undergraduates consisted of 6 EET and 2 MET students. This reflects a broad interdisciplinary interest in this topic.

COURSE STRUCTURE AND DELIVERY

The title of the course offered at ASU East EET/MET 494/598: Fuel Cells: Their Applied Science & Engineering. The course was specifically developed as an introductory graduate level course for students in engineering, engineering technology or the physical sciences. However, advanced undergraduates were allowed into the course subject to instructor approval.
The course was a 4-part, 3-Instructor course, divided as follows:Part 1 Fuel Cell fundamentals and operating principles

- Part 2 Power conditioning
- Part 3 Fuel processing and balance of plant
- Part 4 Laboratory practicals

While the course briefly covered the current state of the art in fuel cell technology, including solid oxide, molten carbonate, Phosphoric acid, Alkaline, and Direct Methanol technologies, the thrust of the course was on the design and performance of Proton Exchange Membrane (PEM) fuel cell systems.

To place the role of fuel cells into proper perspective, the first lecture, given by Dr. Tamizh-Mani, discussed the current state of the fuel cell industry and infrastructure, comparing and contrasting fuel cell technology with existing, proven energy conversion technologies. The goal of this lecture was to give the students a realistic view of existing technology performance capabilities in terms of efficiency, pollution and cost, against which the fuel cell technology must compete.

The next series of lectures, also presented by Dr. Tamizh-Mani, covered the fundamentals of fuel cell physics and chemistry, as well as the technology for manufacturing, assembling and testing assembling the fuel cell stack, with an emphasis on Hydrogen fueled PEM fuel cells. The lectures were designed to answer the following specific questions:

- How does a fuel cell generate electricity from hydrogen and oxygen molecules?
- How much hydrogen or oxygen consumed and water produced during the operation of a H_2/O_2 fuel cell or a stack?
- Maximum theoretical voltage, current and efficiency are dictated by what?
- How to calculate the maximum theoretical voltage, efficiency and current?
- How to calculate the efficiency of a practical fuel cell?
- How to compare the fuel cell efficiency with heat engine efficiency?
- How to theoretically predict the voltage of a fuel cell at a particular current density, temperature and pressure?
- How much heat is rejected (because of thermodynamics and of inefficient electrochemical processes)?
- Can the rejected heat be collected to do some useful work? (CHP, CHCP, Hybrid)
- How much PRACTICAL voltage gain occurs due to an increase in pressure or temperature?
- How long a fuel cell stack lasts?
- What are the construction materials involved in each technology?
- What are the merits and the issues in each fuel cell technology?

At the conclusion of the lectures by Dr. Tamizh-Mani, the students had a fundamental understanding of the design of the fuel cell stack. However, many of the most significant barriers to large scale use of fuel cell technology involve the support systems necessary for fuel cell operation. This includes the control of the DC power produced by the stack, inversion to AC power of sufficient quality for operation of AC powered devices and for transmission on the power grid, the development of highly efficient electric motors, blowers, compressors and pumps for auxiliary devices which minimize parasitic reductions in efficiency, and the production, transportation and storage of hydrogen. The balance of the lecture portion of the course was devoted to these topics.

Professor Sundararajan specifically addressed the following topics:

- The Physics and Technology of DC Regulation and Voltage Conversion
- The Physics and Technology of DC/AC inversion, including both single phase and three phase AC power.
- The Physics and Technology of electric motors appropriate for fuel cell system applications.

Professor Rogers specifically addressed the following points:

- The Physics and Thermodynamics of gases, gas mixtures and gas-vapor mixtures important in fuel cell operation, including psychometric analysis.
- The Physics and Technology of compressors, blowers, fans and pumps.
- Methods of Hydrogen production and transport.
- Methods of Hydrogen storage.

As mentioned, in addition to the lecture portion of the course, the material was reinforced through a series of laboratory exercises, which culminated in the students fabrication of a membrane electrode assembly, and it's testing as a complete single-cell fuel cell in the Fuel Cell Laboratory (FCL) fuel cell testing station at ASU East. The testing resulted in a complete performance curve for the fuel cell, and reinforced the theory learned earlier in the course. The following figure illustrates the testing apparatus:

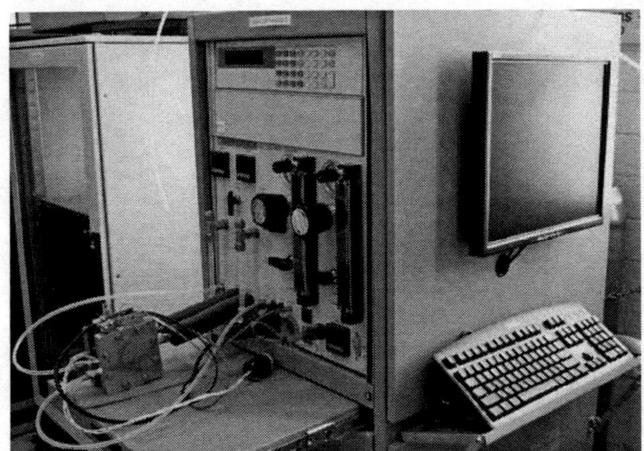

Figure 1. ASU PTL Fuel Cell Testing and Characterization Station

In addition to the manufacture of the membrane electrode assembly, students in the course also made use of the HAAS numerical machining laboratory at ASU East to construct a set of graphite bipolar plates as well as copper end plates. The production of bipolar plates is a difficult machining problem because of the complex flow passages necessary for the efficient diffusion of the reactant gases across the electrode surface. However, the use of modern numerically controlled machine technology allows the production of these plates with relative ease. The bipolar-plates and copper end plates produced in this laboratory are shown in the following figure:

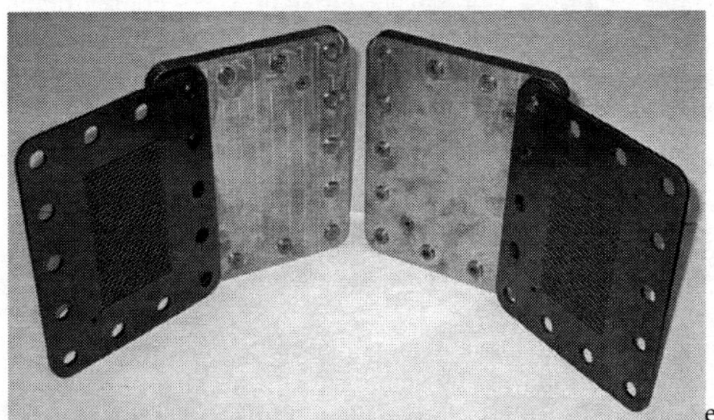

e

Figure 2. Bipolar Plates and End Plates for Fuel Cell produced in the ASU East HAAS Machining Laboratory.

STUDENT PERFORMANCE AND COMMENTS

The students enrolled in this class were generally highly motivated and enthusiastic about the course material. They were very interested in exploring the fuel cell field as a possible career choice, and, in fact, one of the undergraduate students, graduating in the spring of 2003, was employed upon graduation in the fuel cell field based largely on the knowledge gained in this course. At least three of the graduate students have decided to pursue thesis topics in the fuel cell area. Several other students have expressed interest in continuing to explore the field.

The ASU course evaluation process includes an anonymous student course evaluation, as well as an opportunity for comments. The course evaluation is broken in to two parts, that of instructor quality and course quality, each on a five point scale with 5 being excellent. In this case, the graduate students rated the instructors as 4.98/5, and the course as 4.74/5. Undergraduates rated the course as 3.71/5 and the instructors as 5/5. (Department averages for all courses are 4.39/5 for the instructor and 4.27/5 for the course.) Based on personal conversations with students, it is felt that the lower rating by the undergraduates is due to a mild bit of frustration on the part of a few students when faced with some of the more advanced topics.

Anonymous comments obtained through the course evaluation process were highly positive. Specific comments include The course was great all around" "The instructors had a great approach," and "the lab was highly informative."

This feedback reinforces our belief that the course has been a very successful undertaking, and, as will be explained in the following section, plans are underway to expand this curriculum.

FUTURE PLANS

The success of the introductory fuel cell course has led to the planning of a Master of Science program with a concentration in alternative energy technologies within the College of Technology and Applied Sciences at ASU East. This program will be interdisciplinary, involving faculty from the EET and MMET departments, as well as researchers from the Fuel Cell Laboratory. The program will seek to attract qualified students with backgrounds in Engineering, Engineering Technology, Physical Science and Mathematics who are interested in an applied, hands on experience at the graduate level.

To accommodate the program, the course described in this paper will be followed by additional courses in specific topics covered briefly in the introductory courses, appropriate for students from different backgrounds. Several graduate courses appropriate to the field are already in existence at ASU East, including Photovoltaics, Batteries and Electrolysis, Thermodynamics, Heat Transfer, and Fluid Mechanics. Other courses in the planning stages include Wind Energy, and Microturbines. Students will assemble a program of study based on core requirements for the degree, supplemented by specialization appropriate to their backgrounds and interest. The planning for the curriculum is proceeding, and is expected to be complete in the Spring of 2004.

REFERENCES

[1] Larminie, J. and Dicks, A, 2000, *Fuel Cell Systems Explained*, Wiley

[2] Bush, G. W. United States State of the Union Address, January 2003, http://www.whitehouse.gov/news/releases/2003 (Accessed August 11, 2003.)

[3] United State Fuel Cell Council, March 2002, Fuel Cells in Transportation, http://www.usfcc.com/Fuel_Cells_in_Transportation.pdf (Accessed August 11, 2003.)

[4] United State Fuel Cell Council, March 2002, Fuel Cells: Residential Applications, http://www.usfcc.com/Fuel_Cells_for_Houses.pdf (Accessed August 11, 2003.)

AUTHOR INDEX

MET-Vol. 3
Innovations and Applied Research in Mechanical Engineering Technology — 2003

Book Number: I00691